MW00592409

Natural Acts

Natural Acts

Reconnecting with Nature
to Recover Community, Spirit, and Self

Amy E. Dean
author of *Daybreak: 52 Things Nature Teaches Us*

M. Evans and Company, Inc.
New York

M. Evans and Company, Inc.
216 East 49th Street
New York, New York 10017

Library of Congress Cataloging-in-Publication Data

Dean, Amy
  Natural acts : reconnecting with nature to recover community,
spirit, and self / Amy E. Dean. — 1st ed.
     p.   cm.
  Includes bibliographical references.
  ISBN 0-87131-821-0
  1. Philosophy of nature.   2. Environmentalism.   3. Conduct of
life   I. Title.
  BD581.D278   1997
  304.2—dc21   97-2199

Book design by Annemarie Redmond

Manufactured in the United States of America

First Edition

9 8 7 6 5 4 3 2 1

Contents

Dedication

To every person who believes how strange and wonderful this place called Home is—Planet Earth—with its granite, steadfast, strong, eternal elements; its soft, gentle, delicate, hidden, tender undersides; its silent gifts it bestows upon all living things; its ability to burst with fullness and yet which can sometimes be *this close* to emptiness, but which always responds joyously and miraculously to even the slightest attention and the most tentative gesture of friendship.

"...And He said: This is a beautiful world that I have given you. Take good care of it; do not ruin it."

—Jewish prayer

Introduction

I sat on the splintery bleacher with the hot sun beating down, shoulder to shoulder with sweating strangers on either side of me, mesmerized by what I was witnessing. Nearly a dozen wolves were several feet in front of me; I was separated from them by just a chain link fence. The guide who was leading the tour at Wolf Hollow in Ipswich, Massachusetts, was talking about these wolves, which were being raised in captivity to enhance awareness of the creatures and their behaviors. I listened to the wealth of information she was sharing, but never took my eyes from the wolves. I watched them walk around, sprawl on the ground, scratch, pant, romp with one of their keepers, eat, drink. But what moved me the most was what happened after the tour guide said she was going to try to get the wolves to sing. She cautioned us, "They may not want to sing today because it's so warm," then tipped back her head and emitted quite a good imitation of a wolf howl. I stared at the wolves and wondered how they would respond to this human attempt to replicate their wild song. At first, a gentle ripple seemed to pass over the wolves as they listened to her, as if a pebble had been thrown into a still pond. Then, one by one, in careful and precise motion, the pack shifted like choir members who have been cued to move to their rehearsed positions prior to singing their selection. They stopped, then slowly raised their noses to the sky. One wolf began its primal song; another joined in with its singular voice, then another and another, until the pack was united in vocalizing its mournful song. Overcome with emotion, I closed my eyes and let their call reach somewhere deep inside me, deep into a heritage that linked me with the wolf, deep into a sensory awareness that lay buried somewhere beyond my ability to see, to hear, to smell, to taste, and to touch. Tears flowed from the corners of my eyes; I wept for the beauty of the wolves, the glory of their song, the joy of their presence. In that moment, the wolves were not just living creatures that touched me with their presence and, as well, moved me to their plight. The wolves and I became one. I gave myself up to the wolves and, in so doing, experienced incredible joy, deep gratitude, and a sense of silent partnership with them.

"For we are no longer isolated, standing like starry visitors on a mountain-top, surveying life from the outside," wrote Argentine-born English writer W. H. Hudson, "but are on a level with and part and parcel of it; and if the mystery of life daily deepens, it is because we view it more closely and with clearer vision." This book of nature is intended to pull you out of your isolation and encourage you to seek a newer, more natural world in which to live—a world you create for yourself through natural actions you take because you wish to view your world more closely and with a clearer vision. This "more natural" world is one beyond sunsets and roses and sandy beaches, beyond full moons and deserts and mountain peaks, beyond black-capped chickadees and black bears, beyond the natural world as you currently know it. It is, instead, a world of "second nature" in which you are given the opportunity to see and experience the natural world in a different way, perhaps with a different point of view or opinion, and, because of this, are thereby challenged to shift from merely being an observer and appreciator of life and all living things to being an equal partner and vital participant in the day-to-day changes and growth in the natural world.

You live in a world today that is quite different from that of your ancestors, who lived close to the land, who gained their knowledge and wisdom from the natural world, and who moved in harmony with the natural cycles. Every day you absorb enormous amounts of information, sacrifice more and more of your time and attention, and overextend yourself physically, emotionally, and spiritually. For you, as well as for the majority of the population, natural tranquility has become so thoroughly disrupted and destroyed that your inherent natural senses now lie buried far beneath mountains of material possessions, hills of unhealthy habits, and lofty peaks of emotional stress and physical distress. Further, division and destruction of the environment as well as contradictory and confusing information supplied by environmental impact studies force your isolation and disconnection from nature—whether in your own backyard or in the back country. Not knowing whose information to trust, which organizations to back, or who to believe has rendered you helpless and therefore more willing to give up your vote, your opinion, your voice, and your participation in the meaningful and rewarding struggle for the preservation of the natural world.

This book is not intended to be a primer on recycling or activism, a promoter of one or two particular causes, or yet another depressing look at the dire state of Planet Earth. Rather, this book desires to raise your awareness about how to form a relationship with nature that can help you to connect or reconnect to your natural community, your inner spirit, and your self. Nature has always been and will continue to be the living source of everything you are and the integral element needed to ensure your survival. You are part of this intricate web nature has woven over an incomprehensible amount of time. Once all things natural used to fit into your world as the answer to life; your tongue evolved with the taste of the apple, your skin with the dangers that threatened it. Whatever was not needed by you nature transformed into something that contributed to ongoing life in meaningful ways. Nature took care of you, nature took care of itself, and, in return, nature was honored, respected, and protected.

You reconnect with this natural world when you heighten your awareness of your relationship with nature through understanding that, in reconnecting with nature in any one or more of the ten ways presented in this book, you reform the community in which you live in positive ways, redevelop a profound ecospiritual connection, and rediscover your self through learning how you physically and emotionally integrate with nature. In setting forth on this journey to discover or rediscover nature in deeper and more meaningful ways, the hope is that you will learn that nature is really about simple truths, genuine ideas, joyful adventures, delightful surprises, incredible powers, positive changes, ethical principles, strong emotions, beneficial outcomes, and ultimate challenges, all of which can help you to understand yourself and the world in which you live in very profound ways.

fiddler crab, Provincetown, MA, 1956 © David Vestal 1997

In Search of the Simple Life:
Restoring Natural Order

"The ultimate source of the Susquehanna River was a kind of meadow in which nothing happened: no cattle, no mysteriously gushing water, merely the slow accumulation of moisture from many unseen and unimportant sources, the gathering of dew, so to speak, the beginning, the unspectacular congregation of nothingness, the origin of purpose.

And where the moisture stood, sharp rays of bright sunlight were reflected back until the whole area seemed golden, and hallowed, as if here life itself were beginning.

This is how everything begins— the mountains, the oceans, life itself. A slow accumulation—the gathering together of meaning."
—American writer James A. Michener, from *Chesapeake*

Rarely in American life today is there ever "a slow accumulation." Right from birth, material possessions proliferate; one stuffed animal can breed a litter of six by baby's first burp. Each succeeding year welcomes more and more *things* that fill every available nook and cranny. Soon the menagerie of stuffed animals gives way to a toy store aisle's worth of toys and games, which gives way to a locker full of sports equipment, which gives way to a gallery's worth of art supplies, which gives way to shelf after shelf of trophies and ribbons, which gives way to something else, which gives way to something else, and so on. Like insulating foam sprayed inside walls to swell and fill and plug every hole, so too do personal possessions ooze into every available space. Your child grows into Oscar Madison, buried somewhere in his or her room under piles of stuff. *Lots of stuff.*

This same stuff has followed you everywhere and into each of your succeeding years. Eventually you reach adulthood. Your closets burst with enormous collections of clothing and rugby-style scrummages of shoes. Your kitchen counters are cluttered with the latest culinary gadgetry while your refrigerator houses interesting—but inedible—biological growths. Your garage is no longer a repository for your car and car accessories but functions as your home warehouse, filled from floor to ceiling with "overstock." Your television broadcasts the four major networks—as well as over one hundred other channels. Piles of magazines, catalogs, and unread daily newspapers tower over you like redwoods as you sit in your favorite chair in your living room. Your repeated calls to home and yard services, day-care centers, and pet sitters go unanswered, for they already have long waiting lists. Your social calendar is crammed like a physician's appointment book; your household runs as far behind as the average doctor's office. Your children are swept away on the fast track as rapidly as you are; twelve-hour days are not unusual as your children struggle to stay afloat on a whole raft of activities—ballet lessons, religion classes, play dates, arts camp, sports camp, Little League, swimming lessons, scouting programs, birthday parties, and sleepovers, in addition to attending school and completing homework. (A clinical psychologist once summed up the impact such pressures have on children by recounting what happened as he was observing a four-year-old in preschool. He asked the child what she was doing; the little girl replied, "Can't talk now. Working on *Workbook 2.* Going to *Workbook 3.*")

The cultural pressure you feel as well as teach your own children to feel is not just to consume but to gorge: to swallow without chewing, to rest without sleeping, to read without understanding, to look without seeing, to recreate without playing, to connect without bonding, to orgasm without intimacy. The pressure is to enjoy all the advantages, grasp all the opportunities, and experience as much as possible without setting any limits.

When, in April 1996, seven-year-old Jessica Dubroff realized her dream to fly—a dream that turned into an ugly nightmare from which she never awakened—people heatedly began to question just how far parents were pushing their kids to set and achieve tremendous goals, with the sky being the only limit. Yet no one seemed to object to Jessica's lofty goal until after the plane had nose-dived shortly after take-off from the Cheyenne airport. Then, overzealous parents, a maniacal media drawn to the unique human-interest story, and the Federal Aviation Administration, which ultimately permitted the child to fly, were the scapegoats who had pushed the plucky, pug-nosed girl to reach new horizons and achieve new heights. But had Jessica safely landed her Cessna 1778 in Falmouth, Massachusetts as planned, she would have been proclaimed a heroine and her parents viewed as role models for nurturing such freedom and independence in their child. She would have embarked on a whirlwind cross-country tour, hit all the talk shows, been signed to a multimillion dollar book deal, licensed her name and likeness on clothing, endorsed a few products, and been pressed to set a new and more daring goal.

Had this much hype surrounded Henry David Thoreau's unique and successful journey into solitude at Walden Pond, today everyone would be living in huts in the woods and striving for, in Thoreau's words, "simplicity, simplicity, simplicity." Yet nothing in human life is simple today.

Pressed for time and overloaded with gadgets, you, like many Americans, probably feel as if your life is spinning out of control much as Jessica's airplane did. Safe landings no longer are assured; nearly every advance and every advantage in life has its negative, its down side. The coveted MBA might as well stand for Master of Busyness and Aggravation, for along with the degree comes the guarantee of longer working hours, a bigger home, perhaps a "small" vacation home to maintain as well, two cars, maybe a boat, a health club membership, and numerous other accoutrements that must be proudly displayed like medals—medals earned on the raging battlefield of life.

Yet American life just a century ago was impressively different from today. Step outside during the day, pause for a moment, and simply listen. Background traffic noises, overhead jet planes, the wails of passing sirens, the steady moans of leaf blowers or the deafening roars of chainsaws and motorcycles, the whining of construction saws, and numerous other noises continually assail your senses. Step back in time, however, and you will hear the gentle songs of birds, the chirrups of crickets, the soft whispers of winds sharing secrets with the trees, the ecstatic bubbling and babbling of brooks, the steady clop-clop of horses as they trot by, the creaking of rocking chairs as they mark time on front porches, the friendly yoo-hoos hollered from across the street. The only man-made sounds Thoreau heard from his crude shack on the shores of Walden Pond during his period of solitude were the periodic rumblings from passing trains on nearby railroad tracks or the occasional rattling of carriages on a distant road. Today, perpetually traffic-clogged Route 2—the main suburban link with Boston—intermingles the whine of tractor-trailer trucks and the constant whoosh-whoosh of rushing cars with the sounds of robins, blue jays, and mocking-birds; not too far away, Hanscom Field contributes its own offensive aircraft sounds. Lest you think a journey to the wide open spaces of America's West might help you to reexperience the great and pervasive quiet naturalists once reveled in as they paddled the Colorado River through the Grand Canyon, think again. There are now 90,000 commercial tour flights per year over the Grand Canyon National Park; no hiker or rafter has yet to report a trip that was entirely free from the invasive thwap-thwap-thwap of a helicopter. The Great Smoky Mountains National Park, which straddles the Tennessee-North Carolina border and is the country's most visited national park, also has its share of helicopters that hover above hikers whose purpose, in going into the mountains, is for the peace and quiet of nature. Such incessant auditory invasions of the unnatural kind in the natural world has led Dennis

Brownridge, an environmental activist in Arizona, to remark, "There isn't any place I know of that sounds like it did even fifty years ago. And it has crept up on us very insidiously, without our even realizing it."

Too, the night sky looks quite different than it once did fifty years ago. Venture outdoors on a clear night and look up at the sky; sadly, the closer you live to a city, the less you will be able to see. Of the 2,000 to 2,500 stars that are visible over North America in the summer, only a few hundred can be seen from most cities and towns. Astronomers estimate that only one American in ten now lives in view of the full nighttime glory of the universe. City stargazers can no longer appreciate the night skies as they did a century ago; it is now impossible to see the Milky Way or the thousands of faint stars that dot the heavens through the haze of the city's bright lights. This so-called "sky glow" from highway construction, light pollution, and suburbanization—like noise pollution and daily stresses and pressures—deprives you of natural escapes—even tiny sojourns into nature—in which you can take a few moments to appreciate nature's gentle orderliness in order to soothe and calm your soul. "My life moves along at a frenetic pace," writes Andrea Carlisle in the *Chicago Tribune Magazine*. "Just this morning I sent two faxes, responded to seven messages left on my voice-mail, and worked on the computer. At midmorning, however, when clouds gathered and a chill came over the river, I stopped everything to build a fire against the cold. And building it did just what I wanted it to do—slow me down."

How often do you take time to build a fire—either realistically or metaphorically? Do you even have time to exert the effort to create a woodpile from wood you have been able to scavenge or purchase? When was the last time you swung an ax over a chunk of wood and heard the satisfying crack as the wood split? For that matter, have you ever swung an ax? How about a hatchet?

The next time you hear the machine-gun staccato of a pileated woodpecker, try to find the bird. Watch the woodpecker at work on a tree. First it cocks its head to listen for an insect grub chewing a tunnel through the wood. Then it starts chopping a hole near the sound: first in one direction, then the other, just as a person does with a hatchet. Chop-chop-chop, chop-chop-chop. So much searching, so much hammering. Then the woodpecker stops, stretches its sticky tongue deep into the hole and extends it into the tunnel, makes contact with the grub, and pulls its tongue back. A lot of effort, a small reward, but the end result is fuel for the woodpecker to burn so it can stay warm as it searches for its next morsel.

A familiar proverb tells you, *"Chop wood, and it warms you twice."* That's a deal even better than double manufacturer coupons. But how often do you take advantage of this? Perhaps you just get up from the couch during a commercial break and turn a knob on a wall for more heat. Or maybe you use a programmable timer that turns your heat up and down at set times, thereby freeing up a few precious moments you can devote to another task. Whichever way you bring heat into your home during a cold spell, you can still think of yourself as the keeper of your own flame. But how you fulfill this role can either add a rich, meaningful perspective to your life or be viewed as a sacrifice or a burden.

"What is wrong with us?" asked *Cleveland Press* editor Louis Seltzer in an editorial written in the summer of 1952. "It is in the air we breathe. The things we do. The things we say. Our books. Our papers. Our theater. Our movies. Our radio and television. The way we behave. The interests we have. The values we fix.

"We are, on the average," Seltzer continued, "rich beyond the dreams of kings of old. Yet something is not there that should be—something we once had." Forty-one American newspapers and magazines reprinted his words at the time; letters and phone calls flooded his office to tell him he had put his finger right on the crux of the problem. Yet although Seltzer may have been adept at identifying the problem—or even in rallying others to his point of view—the problem was never resolved.

Time magazine essayist Jeff Greenfield opines today, "It is finally dawning on us that we may have made a Faustian bargain a half-century ago, swapping community and neighborhood and roots for the expectation of material abundance for ourselves and our kids, only to find stagnant living standards and overworked two- and three-career families making that promise more and more dubious for more and more of us." Greenfield

Coney Island, NY, 1956 © David Vestal 1997

has, like Seltzer, put his finger right on the crux of the problem today. But what is the solution? What can save Americans from the desire for more? What can ease the stress of daily life? The job pressures? The crammed social calendars? The closets and drawers jammed with useless stuff or items with price tags still on them? The kids' schedules? The junk mail? The endless e-mail and faxes, recording machine messages, the channel changer that might as well teach children to count up to eighty?

One solution to try: *Restore natural order.* Natural order is the way of things, not the way of doing things, the way things ought to be, the way you want things to be, the way others have or do their things, or even the new and improved way of doing things. Natural order is *the way*—the simplest way—the way things naturally occur in the natural world and have done so since creation despite, or, perhaps, in spite of, all the man-made obstacles and hin-drances that have impacted upon it. The natural order is for the sun to set, and then to rise. The natural order is for the tide to come in, and then to go out. The natural order is birth, growth, decline, and death; spring, summer, fall, winter. The natural order is animals coming and then animals going—migrating, hibernating, stalking, killing, mating, parenting, nesting, denning, building, swimming, flying, climbing, tunneling, burrowing, scratching, clawing, slithering, shedding, molting, preening, displaying, marking, protecting, collecting, storing, pollinating, floating, paddling, hopping, leaping, breaching—naturally going about their way, passing and coming. Although the natural order can be disrupted by nature's own hand, such as by a natural disaster or a change in the weather, as well as by man, nature holds true to its natural order and adapts to its ever-changing environment. Nesting birds, for example, have had to become pretty creative over the years in finding suitable locations for their nests because of the loss of forests—the overhang in front of the K-Mart will do in a pinch—and the subsequent loss of forest floor nesting materials. There are true stories of such inventive nests: a robin once incorporated a ten-dollar bill into her nest and one particularly creative and determined European redstart used in her nest, "… 361 stones, 15 nails, 146 pieces of bark, 14 bamboo splinters, 3 pieces of tin, 35 pieces of adhesive tape, 103 pieces of hard dirt, several rags and bones, 1 piece of glass, 4 pieces of inner tubes, and last, but not least, 30 pieces of horse manure!"

Shake a throw rug outdoors or toss a few short pieces of yarn under a nearby tree in the late winter, and, if you're able to later examine any nest that has been subsequently built, you will most likely find strands of your own hair, your pet's hair, lint and threads from your clothing, and the pieces of yarn you made avail-able woven into the nest. Birds could care less whether their nests are lined with spider-egg casings and bark fibers or Christmas tree icicles and whatever drops out of your wallet. They need to stay warm and provide a comfortable nest for themselves and their young, and they will do it in the simplest way, for that is the natural order of things.

Restoring a similar natural order in your life requires that you emit just one word, preferably in a loud and frustrated voice: "Enough!" When you have reached that limit—and you will certainly know when you have because your impulse is to push away from any more excess rather than to crave more—then you are ready to simplify your life.

Simplicity, of course, means different things to different people. To some it means having less stuff and more time to enjoy life. To others it means getting back to basics—making a cake from scratch, walking or bicycling to nearby stores to run errands, writing a letter rather than sending e-mail, playing a board game with the family. Too, there are those who believe simplicity is a philosophical concept that raises vital and thought-provoking questions about the way life ought to be lived; the answers to such questions can lead to radical shifts in the way a person lives—leaving a full-time job to become a self-employed consultant or freelancer, moving away from the city to the country, or choosing to purchase organic food.

But no matter what simplicity means to you, it invariably involves the process—done on either a conscious or a subconscious level—of distinguishing your wants from your needs. In paring down a sweater drawer for donations to a charitable organization, for example, you probably create two piles: one for the sweaters you want, and one for the sweaters you no longer want. But distinguishing your wants from your needs necessitates the creation of one more pile. This pile is for the sweaters you need. *Really need.* Think about it. How many sweaters *do* you really need? How many pairs of socks? How many coats? How many hats? How many dishes? How many cars? How many pairs of shoes? How many credit cards? How many CDs? How many bathing suits? How many towels? How many hobbies? How many collections?

Actress Delta Burke once confessed in an interview with Rosie O'Donnell on *The Rosie O'Donnell Show* that she has a bread roll collection. She said that every time she goes out to eat, she brings home a bread roll, stores it in her freezer, and then, when she has time, she shellacks the roll and adds it to her collection. In fact, she laughed, she shellacks everything she wants to hold onto as a memento—slices of birthday cake, Easter eggs, Christmas cookies, and so on. "It is amazing the things people collect," remarks Barbara Grizzuti Harrison in *An Accidental Autobiography*, "toupees, cookie jars, trolley-car tickets, fans, forceps, dogs, canes, Presley stuff, hatpins, forged signatures, first editions, and gas masks. A Russian countess paid extravagant sums for bedpans that had belonged to famous or notorious people." Only human beings live on such a want level; all other living things live simply on a need level. Creatures eat to satisfy hunger; sleep when tired; mate to ensure survival of the species; relocate to new shelters as needed; have no possessions. *Not even a sweater.* While it may be true that some animals hoard food, they do so not because they want all that food but because they foresee a need for food in the future. And while it may also be true that some creatures, such as ravens, collect shiny objects, they do so not because they wish to have more shiny objects than other ravens or because they plan to trade the shiny objects later on for something else, but simply because the objects are attractive to them.

What can you do without? Not water. Not sleep. Not food. Not shelter. Not clothing. These are your basic needs, similar to the needs of every other living creature on earth. You need to stay hydrated, get your rest, eat, keep warm and dry. However, you differ from all other living creatures because such things cost money, so you also need a job or another source of income. Too, because, you are an emotional and relational creature, you also need love and companionship to combat your isolation and loneliness.

As an individual, you may have countless other needs. Perhaps you need to live in the city because of professional opportunities. Maybe you need to live in a house rather than an apartment because you own pets. Perhaps you need medication because of an illness or chronic condition. Maybe you need to sleep during the day because you work at night. Perhaps you need a wheelchair because you have a physical disability.

What do you need? If this is too difficult to determine, what can you do without? Take stock of your possessions, your lifestyle, your nifty gadgets, your tools, your household decorations, your hobbies and collections. Think of simplifying your life in terms of reduction; let Henry David Thoreau offer good, sound advice to guide you: "Let your affairs be as two or three, and not a hundred or a thousand; instead of a million count half a dozen, and keep your accounts on a thumbnail. In the midst of this clapping sea of civilized life, such are the clouds and storms and quicksands and thousand-and-one items to be allowed for, that a man has to live, if he would not flounder and go to the bottom and not make his port at all, by dead reckoning, and he must be a great calculator indeed who succeeds. Simplify, simplify. Instead of three meals a day, if it be necessary eat but one; instead of a hundred dishes, five; and reduce other things in proportion."

In any search for the simple life today, one argument that is always presented is how the world is radically different than it was a century ago, fifty years ago, even twenty years ago. "Things just aren't as simple today as they were yesterday," is the frequent comment, and you would probably agree. You live in a "too-much" world—a world of excesses as well as extraordinary highs and lows. You can be a computer engineer one day who is making over $100,000 a year plus stock options, paid sabbaticals, an anything-goes work wardrobe, and flex hours, and then suddenly, the very next day, you can be out of a job, collecting unemployment and falling deeper and deeper in debt. You live in a time of high risks and big losses or big gains. You live in a time when the best offense is aggression, assertiveness, and selfishness. You live in a time in which apprenticeship, mentoring, and long years of study and preparation are considered to be a waste of time. "Jump right in!" you are urged. "Get your feet wet!" "You'll never know unless you try!" "Go for it!" and "Just do it!" are some of the rallying cries designed to coax you away from diligence, insistence, research, long hours of study, the search for knowledge, and the pursuit of perfection. Unless the reward can be immediately earned, the consensus seems to be, "Don't even bother." In the race between the tortoise and the hare, today's money is definitely on the hare.

In the natural world, persistence and patience are the two things needed to ensure survival. In the winter, for example, the coyotes at Yellowstone National Park have a particularly hard time finding food. They grow lean as they slowly lose body fat during the cold months; their staple diet of rodents—pocket gophers, ground squirrels, and voles—all but disappear. Even an ailing elk can be nearly impossible for a lone coyote to snag. But the Yellowstone coyotes have numbers on their side and have learned to use patience skillfully as their weapon. A park ranger once observed a pack of coyotes which, for seven days, beseiged an elk that had taken refuge in a stream. The pack finally killed the elk when fatigue drove it to shore. In another example of patience and persistence, wildlife photographer Michio Hoshino, who spent years in Alaska's Denali National Park Sanctuary before he died, captured on film an incredible encounter between a mother moose and a grizzly bear. The grizzly caught scent of the cow's calf and gave chase to it, forcing it into a raging river, where the bear certainly would have killed the calf. But suddenly the mother moose charged out of the brush and into the river at the grizzly, strategically placing herself between the calf and the bear. The cow then reversed the predator-prey roles and charged at the bear. Shrinking back from the enraged moose, the bear turned and fled. But the cow was not about to give up. She continued in her adversarial role, chasing the bear back across the river and up a nearby mountain. Toward the top of the peak, the mother moose seemed satisfied that she had firmly made her point to the bear and that the danger had passed. She trotted back down to the river, but then could not locate her calf, who was terrified from its near-fatal encounter and in hiding. The moose let out a roar and once again charged back across the river and up the mountain, catching up to the bear and confronting it once again, as if to reconfirm that the grizzly had not taken her missing calf.

It has been said that "To make a peach you need a winter, a summer, an autumn and a bee, so many nights and days and sun and rain, petals rosy with pollen—all that your mouth may know a few minutes of pleasure." Years ago, it was more common to be so naturally persistent and patient, to appreciate how things grew over time, and to value long-term investments—those things that

eventually paid off. Nest eggs, saving for a rainy day, putting things aside for a "hope chest," saving up a weekly allowance, investing in U.S. Savings Bonds, contributing to a Christmas Club, and collecting Green Stamps, box tops, or bubble gum wrappers were some of the ways people took their time to get what they eventually wanted. Even though the "big payoff" sometimes brought nothing more than "...an intense and fleeting joy, the final affirmation of a dream, ... [t]he deeper satisfaction," writes Ben Watson in *Yankee* magazine, "was always in the growth itself, the greening force that carries all of the sweet, uncertain hopes for the future."

Look back on your own childhood. Remember how you could roam through your safe neighborhood and play games with your friends without the need for fancy equipment? Sometimes all you needed was a warm summer night and a clean glass jar in which to collect fireflies for hours of entertainment. A comic book or a transistor radio were your two biggest distractions from studying or doing your chores around the house; the television was usually not turned on until homework was completed and dinner dishes washed and put away. The working parent might toil long hours at the office, but life would never revolve around the office so that taking care of the yard, tinkering with the car in the garage, making a minor home repair, tossing a ball in the back yard, going for a drive, reading the newspaper, sleeping, or socializing would ever take a back seat. Those who work today in the high-tech industry—sometimes ninety- to one hundred-hour work weeks—and who often tote cellular phones with them on hikes into the woods, lug laptop computers home at night, move pets into their offices because they are always at work, and rise at four-thirty in the morning because it is the only time to squeeze in a workout know

nothing of such things. They have opted not just for a profession but, in many ways, a lifestyle. Yet it is not a natural lifestyle.

"This is how everything begins," writes James Michener, "—the mountains, the oceans, life itself. A slow accumulation—the gathering together of meaning." You—like the mountains, the oceans, and life itself—have an unknown distance to go in your life from the moment you are born; it may be a few decades or several. You, too, have unknown obstacles you have to face and overcome; they may be life-threatening or merely challenging. You also have twists and turns that will force you into unknown regions, some which you may abhor and some which you may adore. Everything you do, everywhere you go, everything you learn, and everywhere you turn provides you with some level of experiential understanding which, if you can be patient and keep the complexities in your life to a minimum, can gradually lead you to a deeper understanding of yourself, your life, and the world in which you live.

As you live each day, you may conjecture many things, long for great wealth and fame, wish and hope that your dreams come true. You may court doubt and darkness, search endlessly for security, dig deep to find courage, be stubborn in conflict, and be compassionate in all the consequences of living and dying. There is only one world in which you will live and grow, age and decline, and eventually die. This is a world that can either press upon you each minute of each day, snaring you in the stress

Why, AZ, 1966 © David Vestal 1997

of day-to-day living and trapping you in the rat race, or a world in which you experience miracles because you allow yourself to live fully, completely, and appreciatively in each minute of each day.

Ask yourself, "How am I going to live my life?" Think about how you spend each day, then simplify your day-to-day stressors. Improve the environment in which you live. Scale down. Cut down. Cancel. Eliminate. Consolidate. Share. Barter. Slow down. Take your time. Look both ways. Contemplate. Meditate. Recreate. Hold hands. Hug. Create. Choose. Connect. Communicate. Take your time. Take a different path. Try something different. Take a risk. Get outside because "... it is in our nature to go out, to climb the mountains and sail the seas, to fly to the planets, and plunge into the depths of the oceans." Be joyful, for joy is, after all, the purpose of life. Do not live to eat and eat to live. Do not get up each day to

make money and go to bed at night thinking of ways to make more money. Do not accumulate to impress or possess. Instead, slowly accumulate *the meaning*.

Start now to live in the way of the natural world. Revel in those things that take years to mature. Curb impulses; be patient, and patience will be your reward. Enjoy the freedom of living simply, in the natural way. Appreciate your personal freedom. Live life. Enjoy life! As American essayist E. B. White once observed, "No two turtles ever lunched together with the idea of promoting anything. . . . Turtles do not work day and night to perfect explosive devices that wipe out Pacific islands and eventually render turtles sterile. Turtles never use the word 'implementation' or the phrases 'hard core' and 'in the last analysis.' No turtle ever rang another turtle back on the phone. In the last analysis, a turtle, although lacking know-how, knows how to live."

Some Ways To Restore Natural Order in Your Life

1. Become familiar with your night sky. Learn about constellations and the planets. Look through a telescope. Observe a comet or a lunar eclipse. Take an early morning or late evening walk. Support legislation that requires state highway officials to replace roadside streetlights with "full-cutoff" models that prevent light from flowing upward in order to appreciate more fully the nighttime glory of the universe. (Voters have already been successful with statewide light pollution policies in Arizona, New Mexico, Maine, and Connecticut; many municipalities from other states have adopted light control ordinances.)

2. Fill your home or apartment with cedar chips, pine boughs, dried stalks of herbs and spices, or naturally aromatic scents. Bring something from the outdoors into your home—flowers, stones, driftwood, seashells, autumn leaves, a broken bird's egg. Set up a living display of nature in your home with an aquarium, a terrarium, or an ant farm.

3. Support bans or restrictions on helicopters, land recreational vehicles, and motor-powered water equipment such as jet-skis on the national, state, and local level in order to restore "natural quiet" to the countryside, forests, lakes, and seashore.

4. Remember that kids need, want, and benefit from clear, predictable boundaries as well as revel in spontaneous outdoor adventures that involve exercise and play. Let your child be a child before he or she is an adult, no matter how talented, driven, or gifted you believe your child is or could be.

5. Try to sit without moving, without doing anything, and without thinking about anything for longer and longer periods of time. Begin with thirty seconds and then gradually increase the time. Eventually you may find that you are able to calm your tension and restlessness so you

can dally, contemplate, wonder, admire, peruse, marvel, relax, and enjoy the world around you.

6. Desire simplicity in what you eat and how you eat it. Restore the primal pleasures that can be derived from preparing meals without gadgetry. Reconnect with family members by foregoing the usual take-out dinner and preparing one evening meal a week together, with each member contributing a part of the dinner—chopping vegetables for a salad, for instance, grating cheese for a homemade pizza, assembling tacos, slicing apples for a pie. Eat by candlelight, or sit in a circle on the floor and have each person talk about his or her day.

7. Rather than discard stale breads, baked goods, crackers, and soft fruits, feed the birds. Create a simple suet feeder from pine cones by filling them with peanut butter and dried fruit; suspend them from trees. Construct nesting boxes from scraps of wood.

8. Every year go through your closets, garage, and attic and collect all the items you have not used for one year or more. Donate items in good condition to a charitable organization. Discard any recyclable items properly.

9. Reduce the amount of time you spend either at work, taking work home, or networking by one hour per week until you feel you have achieved a more comfortable balance in your life. Use nonwork time to rest, relax, and rejuvenate. Take your dog for a long walk in the woods. Go for a bike ride with your children. Make love with your partner. Watch the sun set. Explore a nearby conservation area, state park or forest, pond or lake, reservoir, botanic garden, or wildlife refuge. Learn about bird watching, botany, or animal tracking, then apply what you learn.

10. Find the upside in any downsizing, even if you lose your job. Nudge yourself back into life. Find serendipity in strife. Unleash a daring creativity. Pursue what you love.

The Lost Language of Villages:
Fostering Natural Neighborhoods

"For most Americans, life does not revolve around legislation passed by Congress or the double talk uttered by Washington bureaucrats. For most of us, life is still helping hands and good neighbors. It is lovingly packed lunch boxes, nighttime prayers, dinners well talked over, hard work, and a little put away for the future.

No government can ever command these things, and no government can ever duplicate them. They are done naturally out of love and a commitment to the future."
—writer Gary L. Bauer,
from *Our Journey Home*

After a long day at their jobs and a tedious commute home, after the mail is taken in, the lights turned on, and the dogs fed and walked, Hudson, New Hampshire residents Kent Baker and his wife head straight for their front porch to relax, smell their flowers and herbs, swat at the evening mosquitoes, marvel at the intermittent sparking of fireflies, and greet neighbors who pass by on their evening walks. "We spend more time on the front porch than we do in the back yard," remarks Kent. "It's a way for people to get to know each other."

Once front porches were a home design staple. Later, the comfort provided by air conditioners and the privacy assured by backyard patios—along with urbanization and the affordability as well as the enticement of yard-maintenance emancipation guaranteed by condominiums and townhouses—made front porches obsolete. Even those who owned homes with spacious front porches kept their porches free from chairs or blocked their steps with plants and decorative arrangements to encourage visitors "to go 'round back."

But today the desire for a front porch is resurging, and not just for aesthetic reasons. No longer viewed as a place to leave muddy shoes, to toss the daily newspaper, or to leave package deliveries, the front porch has become a practical and psychological response to a modern life that not only limits but also discourages human interactions. Front porches also foster a sense of community, restore pride in ownership

Champaign, IL, 1954 © David Vestal 1997

as well as provide safety for the neighborhood, offer gathering places within walking distance for others, and make the streets and neighborhood pedestrian-friendly. Front porches are "ice-breakers"; a wave or a simple "Hello" from a porch-sitter to a passer-by can be the start of a "relationship" in which one neighbor learns to watch out for the other and takes an interest in the welfare of another human being. Front porches are also valuable crime-preventers; porch lights illuminate dark sidewalks or alleyways between homes as well as allow home owners to "keep watch" over the neighborhood so that familiarity is established in ways that call attention to unfamiliar cars and strangers in the neighborhood.

But the desire for a home with a front porch is not the only change taking place in communities across America. More and more, people are seeking alternative ways of getting to work that eliminate the use of cars or overcrowded and stressful public transportation. People bicycle, skateboard, rollerblade, walk, run, rollerski, cross-country ski, or even ride their horses to their jobs not just for the exercise or because, in some cases, such alternative methods of commuting are faster than the train, bus, or car. In reality, it is far easier to crowd onto a city bus or read a book on the commuter rail than it is to arise at a much earlier hour, dress in clothing appropriate to the mode of transportation, pack a change of clothes for work as well as shower supplies, dodge traffic, deal with streets made slick by rain or snow, or maneuver around puddles and potholes, and then change out of sweaty clothing, clean up, and dress for work without being late. Even though participating in such alternative forms of transportation may seem to add more stress to the pressure of needing to arrive to work on time, people who do so report how much they enjoy their own form of creative commuting.

Take, for example, the case of *Boston Globe* employee Richard

Pennington, who rides his horse in the rural community of Ipswich, works in the city of Dorchester, and lives in crowded Chelsea. "One hot summer morning," he writes, "I rode my bicycle to the commuter rail stop in Chelsea. MBTA bike pass in hand, I hoisted the bike onto the eight-sixteen train to Ipswich. By nine, I was pedaling to the barn. Next I saddled my horse, galloped and trotted in the woods for an hour or two, and returned to the barn. I hosed off the horse, turned him out into the pasture, took a sponge bath in the hayloft, and changed from riding breeches to bike shorts. I cycled back to the train, took the one-eighteen train to North Station, and then threaded my way through the Financial District and South Boston traffic, arriving at the *Globe* in Dorchester by three o'clock." Pennington's only complaint after this high-energy commute was that he was thirsty.

So why does Pennington, and countless others, choose such alternative means of getting to work? The desire to journey from home to work without needing to start an engine provides a valuable and nostalgic reminder of what life was like in early American villages when homes, jobs, stores, schools, restaurants, churches, and veterinarians and medical doctors were all close by, when there was a certain rhythm established in each villager's existence that was dictated not by a train or bus schedule but by personal preference, when there was a sense of connectedness that linked a person's home with his or her profession, and when there was a desire to be close to the land in ways that called attention to one's natural surroundings on a daily basis—to the crispness of the air, to the slant of the sunlight on the leaves of trees, to the wind and the weather, to the sounds and the smells, to the wild things growing and the wildlife going about its daily business.

The desire for front porches and participation in unique and personal forms of transportation to and from work communicate quite clearly that people are trying to build a different life for themselves in the impersonal and stress-filled 1990s—a life that is fashioned around living and working in a friendly, personable, caring, and enriching community that naturally unites people physically, emotionally, and spiritually with one another no matter what their differences. This kind of a life creates a valuable "soul" within each person as well as within the community as a whole that can be nourished and nurtured, respected and revered, listened to and communed with much as what occurred in the seventeenth-, eighteenth-, nineteenth-, and even the early twentieth-century villages in New England. "The Founding Fathers thought out the village plans we admire," comments architect David Howard, who has studied ancient European villages as well as American villages from the time of the Colonies. "They thought about this deeply: how rich and poor can live together. No one had ever achieved this equality before. It's a hundred-year thing. A nineteenth-century spectacular. We built these villages without zoning or historic districts. It's a mystery. . . . And it's wonderful." Today, Howard and other architects who share his sentiments on the value of villages and the need for their restoration feel that many people today are trying to create or recreate village living and, at the same time, striving to relearn the "lost language" of villages—a language that suburban development, chemically treated lawns, conscious or subconscious tract housing segregation, and aesthetic rather than useful landscaping have all but silenced.

What draws a village together and makes it work is similar to the biodiversity in ecosystems. A group of people living in a community are all different from one another. In effect, each plays a different role, speaks in his or her own voice, and pursues his or her own particular character's needs. Yet, like a group of actors in an ensemble piece, each must work with one another in this group of diversity. Each must ultimately communicate with and understand the same language. A missed line or miscue can impact the flow of the entire presentation, causing minor disruption; a missing actor can completely upset the entire ensemble and ultimately ruin the presentation.

Nantucket, Massachusetts is a prime example of one of New England's historic villages that continues to thrive today because each individual within the community still speaks the "native" language with which it was founded over a hundred years ago when it evolved as a tight fishing village, formed by a way of life adopted by all those who lived in the village. The cobblestone streets and sidewalks still remain, the businesses still cater to their "regulars" as well as to the tourists, community spirit is still high, holidays are still celebrated with family as well as within the extended family of the community, the small home frontages are still kept clean and well-tended, and the houses of both the rich and the poor are still not only indiscernible from one another from the outside but are also located within five hundred feet of each other—quite different from the housing classicism that takes place today in many locales, where the rich gravitate to and then hide behind gated, secured communities while the poor are located miles away. "The biggest house [in Nantucket] is ten times bigger than the smallest house, one block away," comments Howard. "Yet, you never feel they are drastically different houses." This leads Howard to wonder, "What do these people in gated communities say to their children? [In a village] There's a sense that being an honorable person is more important than being the richest person."

Not only do villages speak a language of equality, honor, and respect for tradition among its residents, but villages also communicate to its residents, through architectural design, that there is accessibility to all residents, that a high density of people and houses is good because it lends a sense of security and safety, that social contact can naturally occur because of the close proximity of homes to streets, that residents can walk to valuable services such as the post office, church, the town hall, the police and fire stations, the grocery store, the coffee shop, the newspaper stand, schools, parks, and so on, and that public space has a greater value than private space. Take out a village's Main Street, the town common, or the meeting house; build the houses further apart, perhaps by putting an acre or two between them; move the houses farther back from the roads instead of keeping their frontage to a minimum; or seek to change what has already been established and which works well; and the village makes no sense. It speaks no

language of connection; it is, in effect, muted—grown as silent as a polluted pond that can no longer support the living things it once sustained. Villages cannot survive in the human world—and biodiverse ecosystems cannot thrive in the natural world—when there is no attention paid to how each living thing is responsible to others in its community, how intertwined each living thing is with others, and how dependent all living things are to one another.

When Michael Pollan, Executive Editor of *Harper's* magazine and prolific writer on gardening and nature, decided to hire excavators to create a backyard pond behind his house in October 1993, he was amazed at how quickly the pond filled with water and life began to teem in and around it. In March, algae began to drift through the water. By the end of the month, the song of the spring peepers came from the pond, soon joined by a chorus of bullfrogs. In May, backswimmers sought meals of insect larvae and whirligig beetles zoomed across the surface. Water striders, aquatic plants, cattails, spiders, and waterfowl used the pond during the warm months, while the tracks of fox, wild turkeys, deer, raccoons, woodchucks, and various species of birds were visible in the snow.

Elated at the success of his pond, Pollan "was happy now to let nature take its course, pleased to have had a hand in the making of this thriving new ecosystem." He brought a jar of its water to a local biology teacher to gain a deeper understanding of the life his pond was sustaining. As the teacher viewed the water, he informed Pollan that his man-made pond would not remain a pond for very long. Left untended, the algae and weeds would continue to grow, die, and then settle to the bottom. This would make the water shallower which, in turn, would let more light into the bottom of the pond so the weeds would redouble. The pond would

gradually turn into a swamp, then a wet spot, and then return to its original state—woodland—unless Pollan intervened and began a rigorous program of pond maintenance. As Pollan listened to the teacher he realized that, "I had naively assumed that, the digging done, I could step back and let my pond take care of itself. Not so."

Perhaps this is how suburban developers and town officials approached the planning and construction of housing tracts. The philosophy of, "If you build it, they will come," presumes that a sense of community will automatically happen as each new family moves into the development and settles into the rituals and routines of their life. But because any community is made up of a variety of different families, each with its own lifestyles, backgrounds, cultures, ethnicities, religions, beliefs, and so on, it becomes vital that all the diverse families in the community learn to share equally in ensuring not only their own survival but every other family's survival in the community in order to promote the health, growth, and longevity of the entire "pond," or community. Remove one family from the community "pond," and the impact is felt in each of the remaining families; thus, each must exist in order to sustain the others. Yet, on the other hand, each, if allowed to proliferate uncontrollably, will strangle the others, leading to the ultimate destruction of the whole community. A community works best when the individual families are committed to the survival not only of themselves, but also of the others in their community. Supporting the positive development of those who are living in a community can take many forms: by giving to others (helping neighbors paint their house, offering child-care assistance, or baking a casserole for a family that recently lost a member), by participating in organizations that benefit others within the community (coaching a soccer team, starting a neighborhood crime watch campaign, or organizing a multi-family yard sale), by attending town meetings and learning about the voting records and intentions of local and state political candidates in order to protect the rights of the people who are living within the community as well as the natural living things that are important to the community, by celebrating in the joys of others as well as reaching out to them in time of need, and so on. Each experience that shows recognition of others in the community is an investment in the community's future; thus, for better or worse, you and each member of the community ultimately share in the quality of life within the community now and in the future.

Today, a generation of Americans has grown up driving to giant shopping centers to get its groceries and other necessities. In suburbia the streets are wide, the sidewalks broad, and the houses are set back far behind a green moat of lawn. Lawn services show up for their roaring weekly ritual of lawn caretaking, spritz the greenery with chemicals, and then leave flagged warning signs that function as invisible fences designed to keep the household residents inside and the outside community away. The majority of outdoor activity takes place in the backyard, hidden behind the house, shielded by a six-foot stockade fence, darkened by large-growth greenery that is often not native to the area. Suburbia creates isolation and forces soliloquies among its residents rather than bonds them together and unites them in a common language.

A community works best when all of its divergent elements are supported equally. In his classic 1969 work, *Town Planning in Frontier America,* John W. Reps wrote that "public welfare was paramount" in eighteenth-century villages. "A rough but workable democracy prevailed.... Rarely since that time have important decisions been made about community planning in America with such harmony." Today a movement called "New Urbanism" is striving to incorporate into modern developments design elements such as front porches that once helped neighborhoods thrive. Founded several years ago by a group of architects from around the country in response to suburban sprawl and urban decay, the New Urbanists are striving to change the way you live by changing the look of what you live in and the streets you live on. Christopher Horning, deputy assistant secretary for the U.S. Department of Housing and Urban Development (HUD) comments, "...design really has an enormous impact on how people live and interact. It's just become so critical that the architecture reinforce what is strong in the community...."

Biodiversity similarly supports all living things within their own "villages," making each strong. Disrupt one element however, and the entire village is disrupted; unless corrected, the village, or habitat, may ultimately be destroyed. Take, for example, what happened in the 1950s when the World Health Organization tried to eliminate malaria in northern Borneo by using the pesticide Dieldrin to kill mosquitoes that were carrying the disease. Almost immediately, the project appeared to be a success. The mosquitoes disappeared; incidences of malaria dramatically decreased. But then something strange happened. The roofs of the villagers began falling in on them and the people faced the threat of a typhoid epidemic. Why? The thatched roofs collapsed because the Dieldrin also killed the wasps and insects that ate the caterpillars that fed on the roofs; hundreds of lizards died from eating the poisoned mosquitoes; the village cats died from eating the poisoned lizards; rats began to run rampant through the village, carrying typhus-infested fleas with them. In this case, two villages were destroyed—both the habitat and the actual village.

A similar thing happened in North America, Europe, and other parts of the world that used organochlorine pesticides, DDT in particular, for controlling insect pests in yards, farms, and forests. While some peregrine falcons that ate contaminated prey died, the greater problem was shell thinning that caused most of the birds' eggs to be crushed during incubation. Scientists eventually determined that DDT's metabolite, DDE, blocked normal calcium deposition during eggshell formation, but by then the noble hawk had been placed on the endangered species list when there were only a few young birds to replace the dwindling adult population. Up until 1970, no peregrine had even been spotted on any cliff east of the Great Plains until observers located a lonely male in Vermont scanning the skies for a mate. As the peregrines moved closer and closer to extinction, the birds' natural habitats began to slowly be decimated, for the birds had lost their voice in keeping their "villages" intact. To save the peregrines, not only did chicks need to be raised in captivity, but new nesting sites—new villages—had to be created for them. In an unlikely working relationship, across America city planners and bird specialists met to set aside suitable sites on buildings and bridges for falcons to raise their young, hunt, and nest. On the door leading to the rooftop ledge of the Washington Mutual Tower in Seattle, for instance, the bank management attached a sign— *Home of the Peregrine Falcons. Please Be Quiet When Passing Through*—to protect the environs of Seattle's "first couple," peregrines Stewart and Virginia, who settled into nest boxes located on the east side ledge. Peregrines now winter and breed in their new, man-made villages in Seattle as well as in San Francisco, Denver, and New York City. "Without realizing it," environmental writer Adrienne Ross notes, "our artifice, in an animal's eyes, recreates the natural world. . . . What we see as a skyscraper of chrome and steel, its windows reflecting clouds in a turquoise sky, a peregrine can see as a cliff, full of crevices for nests and ledges for the clumsy first flights of young. . . . Far from being isolated from nature, our cities are playing an active role in evolution, favoring certain species over others and helping determine what degree of biodiversity can exist into the twenty-first century."

What degree of village living will survive or even be revived in the twenty-first century? In his *Origin of Species,* Charles Darwin stated that natural selection—the adaptation of every organism to a particular way of life—directs evolution. This being the case, what does this say about what kind of people Americans are becoming as the communal aspect of village living gradually becomes a distant memory? In the pursuit of the American Dream, life has, in many ways, probably become a nightmare of adaptations and evolutions for you. Achievements now have value, not people. Happiness now comes from a bigger house or a larger paycheck, not a sense of accomplishment or a smile shared with a loved one. Time is an unaffordable luxury, not just an arbitrary form of measurement or a guideline. Motion is perpetual, civility is forced, individuality is the key.

This American way of life, however, is also the American way of death, for the emphasis is on "only me" from start to finish. In saying good-bye to the close-knit way of life in the village communities of their youth, Americans by the droves have set off on a life of impermanence in pursuit of their own version of the American Dream—moving up corporate ladders, juggling careers with raising children, leap-frogging from academic position to academic position at universities, drifting in and out of relationships and marriages, pulling up roots every few years, living out of boxes, eating from take-out containers. If you barely have enough time to devote to your immediate family, then you probably have little time to give to friends, professional contacts, or members of the community in which you reside. And while you may be well-connected in the professional community and "plugged-in" to an international information network, you may not even be on nodding acquaintance with your next-door neighbor. There should be no excuse for being so emotionally detached from those you see every day, but there are countless ones that are offered. Even though research shows that close ties with others keep you healthier and help you to live longer, often the demands of contemporary life keep you distanced from others until some tragedy—a heart attack, perhaps, or the ending of your marriage—forces you to take a long, hard look at the state of your life. Then, "When each of us must face the fact that our precious time on the planet is almost over, when we reflect on what was most important to us during our brief visit here," Cynthia Dickstein, president of the Organization for International Professional Exchanges, Inc. asks, "will it be our bottom line or our top-ranking position? Or will it be the smiles and the tears and the love of our families and friends who are at our side for our very last good-bye?"

It's no coincidence, for example, that for decades in the sport of competitive running, American runners have rarely won long-distance races such as marathons and, year after year, are rapidly being out-sprinted in some of the more prestigious 10K's, 5-milers, and 3K's. "The American way," writes World Cross Country Championship American runner Joan Nesbit, "is to hole up on our individual runner's islands, month after month, until we've somehow earned the right to race. . . . In the Kenyan system, the youngest (or slowest) runner on the team sacrifices by running in front for as long as possible to allow the designated winner to cruise along effortlessly. When the young Kenyan tires, another takes over to force the pace. The theory is, if you sacrifice your race today, your time will come to be the designated winner. For them, it works. Can you name a single U. S. runner who would be willing to do this for his or her team, or for our country? Sadly, no. It would go against our precious American individualism." The familiar coach's refrain you may have learned in your youth—"There's no 'I' in team"—has long since been replaced by the star-and-supporting-cast syndrome, evident in every major sport today. With young people becoming self-absorbed, the elderly socially ostracized and isolated, and adults living lives in which they run like mice through mazes, are Americans fast becoming a nation of strangers who must learn how to survive in a world with few emotional connections?

McKittrick, CA, 1959 © David Vestal 1997

Without membership in a community, you cannot be held accountable. Without membership in a community, close ties cannot be formed. Without membership in a community, you do not have to be courteous and civil. Without membership in a community, you do not have to care about other people. Without membership in a community, you do not have to be intimate or close to others. Without membership in a community, your survival is in your hands.

In Irian Jaya, the western half of New Guinea in the province of Indonesia, lives a village of tree people known as the Korowai. There, the tribe still lives as their ancestors did tens of thousands of years ago. But soon that may change. The government has targeted Irian Jaya's treasure-trove of natural resources; soon bulldozers will tear down the trees in which the natives live and from which they hunt. Fortune seekers and others who are looking for a better way of life are overcrowding the island under a government-sponsored relocation program and are bringing twentieth-century technology with them, rapidly antiquating tribal ways. The Korowai do not keep count of years and know almost nothing of the outside world, incapable of believing that humans live in "huts" taller than the tallest tree in their rain forest. They know little about medicine; plants and bark are used to heal and bind wounds while rituals take away their pain. And yet the Korowai know that their most powerful spirit— Ginol—the creator of the world in which they live, will not be able to defeat the *laleo,* or the white-skinned ghost demons; they will eventually invade Korowai land and obliterate it.

The Korowai land, the pond, and the village are all communities that thrive because of what happens on the inside, not on the outside. While outside influences certainly have the power to destroy and decimate a community—the Korowai tribe is one of the few remaining rain forest clans; countless ponds now lie buried under layers of dirt and homes, city sidewalks, and industrial parks; and small villages have long since blossomed into metropolises—it becomes difficult to negatively transform such communities the stronger the community is. When you feel that your home is more than just a private location in a town and, instead, feel that you are *at home* in the whole town, then you make a powerful and significant connection with the community. When you feel that you are no longer a spectator who is looking in from the outside at your community, you become a vital participant in the community. You enrich the village's life blood that courses through its veins by giving it life and sustaining that life through actions designed to maintain the integrity and build the strength of the community.

Some Ways to Foster More Natural Neighborhoods

1. Become familiar with your neighborhood and your neighbors. Learn the names of the people and their pets. Compliment household and yard improvements. Share a batch of freshly baked cookies. Shovel out a walkway for a sick, injured, or elderly neighbor. Always smile and wave hello, even when such things may not be reciprocated.

2. Be the type of neighbor you would like to live next to. Keep the volume of your music and television low. Alert those who live nearby about gatherings so you can work out parking arrangements. Teach your power tools to respect late hours during the week and early morning hours on weekends. Keep an eye on your neighbor's property; take down the license number of unfamiliar vehicles. Be thoughtful and generous. Reward your dog for barking only

when a stranger approaches your home. Keep trees and hedges trimmed. Repair sagging fences. Work towards resolving conflicts, not prolonging them. Even when forming a good relationship proves to be difficult, always be courteous and civil. Respect privacy.

3. Attract wildlife to your neighborhood in ways that assist in sustaining the natural biodiversity of the neighborhood as well as maintain a respectful relationship with your neighbors. For example, rather than install annoying bug zappers, burn citronella candles or mount bat houses on your house and the trees around your yard. Secure mulch piles or use enclosed mulch equipment in order not to attract skunks, raccoons, and other garbage-loving wild animals. Hang bird feeders. Plant flowers for hummingbirds, butterflies, and bees.

4. Beautify the front of your home in small ways. Put on a fresh coat of paint. Add colorful flowers to planters. Hang soothing wind chimes and mobiles. Decorate your front door with a natural wreath. Create seasonal and holiday displays.

5. Support a community garden project in which young and old can participate. Share what is grown with one another; donate any surplus to your local food pantry.

6. Who or what lives in your yard? Notice what goes on in your back and front yards. You may see a red-tailed hawk flying overhead, spot a woodchuck or rabbit scouting out the garden, glimpse a pair of lively chipmunks as they scurry in and out of your stonewall, hear the honking of migratory geese, or watch a butterfly delicately sip flower nectar. Create a friendly environs for all living things.

7. Be sensitive to planting pollen-producing greenery that can make it hard for you or your neighbors to breathe during times when the pollen is released. Bermuda grass, cypress trees, and mulberry bushes are a few examples of species that throw pollen

into the air and can make it difficult for those who suffer from asthma or other bronchial disorders to enjoy the outdoors during certain seasonal periods. Also, be aware of the herbs and plants that you cultivate in your yard. Even though they are rooted to one spot, plants defend themselves, stake out territory, lure certain creatures to pollinate them, and repel others.

8. Support local businesses and their small business owners; they keep your community healthy. Work with town officials when business rental space becomes available in securing and welcoming new and exciting businesses into your area.

9. Defend the biodiversity in your neighborhood. Explore the impact any change to the environment will create on other living things. Keep pesticide use in the garden or on the lawn to a minimum; in the winter, use sand, kitty litter, or environmentally safe salt mixtures to melt ice and snow. Explore how to safely and effectively dispose of hazardous household products and liquids such as gasoline, paint, turpentine, motor oil, and so on.

10. Band together with your neighbors to protect and preserve the communal spaces in your neighborhood. Participate in keeping such areas neat and tidy. Help rake leaves, reseed and fertilize grass, plant flowers, and prune and plant trees. Encourage local horticulturists and arborists to provide labels that will identify trees, herbs, and other growth.

Defending a Place: Seeking Natural Preservation

"If the rose at noon has lost the beauty it had at dawn, the beauty it had then was real. Nothing in the world is permanent, and we're foolish when we ask anything to last. But surely we're still more foolish not to take delight in it while we have it."
—American writer W. Somerset Maugham, from *The Razor's Edge*

"I don't know whether a passionate love of the natural can be transmitted or not, but like the love of beauty it is a thing one likes to associate with the scheme of inheritance."
—American writer E. B. White

While virtually all biologists believe that the story of life is, essentially, an unsolvable mystery—for example, it has never been determined whether natural selection alone is enough to explain the diversity and complexity of life on Earth—certain things are held to be true, indisputable "facts of life." Biologists know that life has developed on this planet over the course of billions of years, that all living organisms share common ancestors, that natural selection helps to shape life, and that species come and go. Beyond such facts, however, the story of life is up for grabs; so, too, is its outcome. What happens next depends upon you for, in many ways, you are one of the millions of authors who contribute on a daily basis to this unfinished story of life. Although, as Stuart Kauffman of the Santa Fe Institute in New Mexico writes in his book *Origins of Order*, you cannot "... even begin to understand how selection and self-organization work together to create the splendor of a summer afternoon in an alpine meadow flooded with flowers, insects, worms, soil, other animals, and humans, making our worlds together," you do know how great your impact can be on this idyllic, harmonious scene of biodiversity, for you know now that you have the power to either destroy it or to protect it. You know this because you have either participated, in direct or indirect ways, in the destruction of nature or have experienced the effects of what others have done when they have destroyed a piece of the land on which you live, altered the purity of the water you drink, tampered with the organic qualities of your food, affected the richness of the soil in which you plant your trees, flowers, and vegetables, diverted vital water resources and so parched your land, or vented noxious fumes into the atmosphere that have poisoned the air you breathe.

No longer can you be a silent inhabitant of this planet, choosing to go about your daily business with blinders on and your senses shut off, for it has become nearly impossible not to be drawn into the physical battles that are being waged, the emotional fires that are being ignited, and the spiritual outrage that is being expressed at the destruction, decline, and decimation of Planet Earth and all of its inhabitants. When you see pictures or news footage of thousands of oil-coated sea birds lying dead on a sandy beach after yet another oil spill, the scene becomes gruesomely reminiscent of the beaches of Normandy shortly after the Allied invasion during World War II. The corpses of the dead soldiers—those who had enlisted to defend their freedom, *their right to life on Earth*—that lay strewn about on the beach where they had fallen are little different from the bodies of the sea birds, for both species have died at the hands of opportunists whose desires were profits, their own personal power, and having their own way with the world without regard for the life, liberty, and pursuit of happiness of all living things.

"In a free society every opportunity comes with three obligations," Roberto C. Goizueta once said in a speech. "First, you must *seize* it. You must mold it into a work that brings value to others. Second, you must *live* it. Opportunity is nurtured only by action. Third, you must *defend* the freedom to pursue opportunities. You must embrace these three obligations as if the future of the United States depended on it. In fact, it does."

So, too, does the future of the natural world depend upon your ability to seize, live, and defend it. You must understand all of its creatures, all of its wild growth, and all of its nonliving elements.

You must then translate this understanding into a *value* that can be applied to individual elements and entities—a value that goes beyond your own interests, beyond the interests of your community, beyond the interests of environmentalists and developers, and beyond the interests of corporations or governments *to encompass the interests of all living things*. In your quest for understanding, you must ask not just yourself or the people you know or who share in your sentiments what impact even a small change will make upon an ecosystem. Rather, you must take the next step; you must seize the knowledge and then live it. You must become as actively and passionately involved when just one tree is being taken down in your community as when an entire forest of old-growth trees is slated to be cut, such as in Alaska's magnificent Tongass National Forest. You must learn what shipwrecks, cyanide fishing, harbor dredging, coral mining, deforestation, coastal development, agricultural runoff, and careless sea divers are doing to the ocean's delicate coral ecosystems. You must learn not only where oil comes from, but the price the natural world must pay to extract this oil from the environment's safe holding tanks. You must learn what strip mining does not only to the land but to the nearby rivers, what impact pesticides sprayed on fruits and vegetables have on the human reproductive system as well as on the songbird's reproductive system, how much territory a free-range as well as a zoo buffalo needs in which to roam, how to safely dispose of hazardous household products, how your elected officials vote on environmental issues, what kind of political or corporate deals have been struck in order to protect a wetland, what happens to the fish and wildlife when their water source is diverted for farming irrigation. You must ask questions and you must expect answers to your questions that tell you not

what you want to hear but what you *need* to hear. You must fight to get to the truth and, in all probability, must expend a great deal of your time and energy as well as even endure financial hardship in this process of discovery. And then, after you do all this, you must do one more thing: *you must defend what you believe to be right for the future of the planet.*

You must defend this planet, this place you call home, as fiercely as you would defend your own home, your self, your honor, your children, your partner, your mother, your father, your sister, and your brother. You must defend this place because as you grow older, your natural world grows smaller and smaller and increasingly more sterile, more barren, more silenced. You have the responsibility to teach yourself and others that there is a common ground between the environment and the industrialization of the human mind and psyche. This common ground is *environmental justice,* and it must be vehemently defended for its purpose is to value both human and environmental concerns for the simple reason that one cannot exist without the other.

Defending a place is not often something you set out to do or even choose to do. Defending a place rarely begins as a career goal, although some people have, after striving to make a difference, chosen to devote themselves full-time to the pursuit of environmental justice. While defending a place is rarely something you may ever envision yourself doing, particularly when you are young, you may, as you grow older and more concerned not only about the natural world but those who will live in it after you are gone, end up choosing to do it—or it may end up choosing you.

When Damacio Lopez left his Socorro, New Mexico home to pursue a successful career as a professional golfer, he was enthusiastic about returning to the town where he grew up, to the place where twelve family generations had grown up. He was coming back to Socorro a hero, and there was much joy and celebration in his family upon his arrival. But then he heard a loud explosion near his town that startled him; he watched as a large cloud of black smoke slowly drifted over his home. He asked his family, "What is in that cloud of smoke?" No one knew. His mother told him, "We've gotten used to the explosions even though they sometimes crack our adobe walls." Concerned over the safety and welfare of his family, Lopez stayed in New Mexico for months, later learning that there was uranium in the cloud. It was the mid-1980s, and uranium munitions were regularly being batch-tested by private industry and the New Mexico Institute of Mining and Technology. The ensuing toxic clouds rained radioactivity down upon Lopez's family, friends, and neighbors. "I left my life's work of hitting a ball in the least number of strikes to study the uranium cloud and to learn why people did not ask what was in the cloud," explains Lopez, who abandoned his career as a golfer to become the research director for Revisioning New Mexico.

Gershon Cohen, a former biologist and potter, co-founded Alaska Clean Water Alliance, a state-wide nonprofit organization dedicated to the protection of Alaska's watersheds, after being alerted by a former employee and "eyewitness" about a possible major DDT burial that had occurred decades previous on a decommissioned jet-fuel pumping station just five miles across the border from Alaska in British Columbia. The area, known as Rainy Hollow by Canadians, abuts a critical habitat for nearly four thousand Alaskan bald eagles; leakage of DDT into water supplies would endanger the birds. So Cohen alerted the CBC Radio News department and accompanied an employee from Environment Canada to the site; in less than a week, excavations at the sight revealed forty five-gallon canisters of concentrated DDT that had been buried well below the spring flood water table line as well as several half-crushed and unmarked fifty-five gallon chemical drums. The ensuing million-dollar clean-up led to increased concerns over the illegal disposal of pesticides and chemicals by private industry; Cohen assumed the position of Executive Director of Alaska Clean Water Alliance to protect Alaska's watersheds for the future.

For Penny Newman, Director of the Center for Community Action and Environmental Justice, which works with communities across the nation to deal with their pollution problems, defending a place was the furthest thing from her mind. In 1965, such issues of community empowerment and toxic pollution played no part in her "perfect" existence. She had just married her husband; they had settled into the quaint rural community of Glen Avon, located in a valley between the rolling hills in southern California. There they bought a house and raised their two sons. Penny became a school teacher. She and her husband joined the PTA; the kids joined the Little League and Cub Scouts.

In 1978, after a winter of drenching rains, the ensuing flooding in Glen Avon not only made roads impassable and turned the schoolyard into a lake, but also released a strange, acrid smell into the air. Foam floated on the puddles that Glen Avon's children splashed through; the kids played with the foam, laughing while they made frothy beards on their faces.

"Then," Newman recounts, "frightening things began to happen. Tennis shoes and jeans fell apart; bloody noses were commonplace. Like other children in the community, my boys had red, sore, irritated skin. Shawn developed double-vision and severe headaches.

None of the medicine prescribed would stop his pain. As we searched for answers, our lives changed forever. That winter, without our knowledge, toxic chemicals had been deliberately released into our community." Very few people in the community knew about the Stringfellow Acid Pits, a hazardous waste dump that had been in Glen Avon since 1955, kept hidden in a box canyon less than a mile from the elementary school. The seventeen-acre site held over thirty-four million gallons of discarded chemicals: cleaning solvents, heavy metals, pesticides, acids. The heavy rains in the winter of 1978 created ponds in the holding pits that soon began to overflow; state officials decided to relieve pressure on the dam that was holding back the water in the ponds by releasing one million gallons of the contaminated water directly into Glen Avon— without warning its citizens.

Newman, along with others, formed Concerned Neighbors in Action (CNA) in order to become involved and to help decision-makers with their input in resolving the problem. But rather than be welcomed, the coalition was treated with condescension and indifference; attorneys for the U.S. Department of Justice, the State Attorney General's Office, and the EPA sought to exclude the citizens from their meetings. The CNA refused to give up, defending their right to information as well as to effective resolution of the problem. After seventeen long years of battle, the CNA settled with the polluters for $114 million. "It doesn't make up for the loss of our friends to cancer or the children's repeated surgeries for brain tumors or kidney transplants," says Newman. "But it does provide some vindication

near Highland Light, Cape Cod, MA, 1954 © David Vestal 1997

for the suffering we as a community went through and allows us to rebuild our lives. And it acts as a deterrent."

Newman explains that she and the other defenders of their place never saw themselves as environmentalists but "simply hard-working, good citizens trying to raise our kids and provide a safe, healthy life for our families. Our fight was for our own survival. We came to understand that since it is our community, and our lives, we have the right and the responsibility to actively participate in decisions affecting us."

In defending any place, what is always presumed is that there is an oppressed entity and an oppressor. Sometimes the oppressor is a government, sometimes it's a major corporation, sometimes it's the military, sometimes it's an industry, sometimes it's a community, and sometimes it's just one person. But oftentimes identifying just who is the oppressed and who is the oppressor becomes muddled. For example, does the PETA group— People for the Ethical Treatment of Animals—oppress the rights of those who are sick and dying when it opposes *all* animal medical research? PETA presents little middle ground in the battle between defending the rights of animals and defending the rights of the ailing. On one side of the issue is a vocal host of AIDS activist and research groups as well as prominent non-profit groups that are fighting an array of diseases—from the American Foundation for AIDS Research (AmFAR) to the March of Dimes and the American Cancer Society. Too, there is Jeff Getty, who, in December 1995, underwent a highly publicized baboon bone marrow transplant in the hope that the new cells would battle the HIV in his body. Although the experiment later showed no baboon cells that survived in his body, Getty said that he had experienced "a clinical return of my health to a state that it hasn't been in five years."

PETA was founded in 1980 by animal rights activists Alex Pacheco and Ingrid Newkirk because they felt that the humane treatment of animals, which was being protected by the American Society for the Prevention of Cruelty to Animals (founded in the mid-1880s) and the 1966 federal Animal Welfare Act (which mandated "humane care and treatment" of animals in the United States), was not being adequately defended. "It cannot be right to seek a remedy for HIV through the deliberate infliction of suffering on other sentient species in laboratory experiments," opined one PETA member. While many PETA members continue to believe in the promotion of humane care and treatment of animals through their opposition to fur farms and animals raised for the slaughterhouse, after learning of PETA's stance on animal testing for medical purposes some of PETA's more prominent spokespeople withdrew their support from the organization, including antifur campaigner and rock singer Melissa Ethridge.

Sometimes, in seeking to defend a place, you may be shocked to discover that those who are identified as the oppressors actually feel they are the ones who are being oppressed. Loggers from Maine to Alaska, for example, feel that conservationists oppress them by striving to prevent them from logging trees and therefore pursuing a livelihood. *No Trees, No Jobs!* scream signs of protest against conservationists who seek to stop forest cutting; there are lots of people who are protesting the protestors. Alaska's Ketchikan Pulp Company, a division of Louisiana-Pacific Corporation, employs six hundred people. The dissolving pulp, a cellulose product manufactured by the company from cutting, is used in everything from rayon to ice cream. In the economically challenged regions of northern

Maine, the logging, paper, and pulp industries produce up to $5.5 billion worth of paper and lumber products a year and provide twenty-six thousand jobs. The potential loss of jobs to individuals and the astronomical loss of money to companies—a single Alaskan two hundred-foot Sitka spruce can yield up to ten thousand board feet of timber and is so fine that it can be used to make pianos and guitars, while a California redwood three hundred feet high and eighteen inches in diameter can bring two hundred thousand dollars a sawlog—makes those people and businesses feel justified in believing that their rights, not the rights of the trees, are being violated. Yet the defenders of a redwood tree see more than just a tree that would be lost through cutting. The marbled murrelet—a rare, threatened sea bird that nests only in the top of old-growth redwoods—would lose its habitat, as would the spotted owls and peregrine falcons. Too, coho salmon would be threatened, as would the rivers and streams near clear-cutting, for the waterways would fill with sediment without the trees to keep the forest floor intact.

Because there are always two sides to every issue, and because each side often fervently and unswervingly believes in its own position, the defense of any place must incorporate all angles, all aspects, all considerations. There must be a willingness to not only look at both sides of an issue by the two parties involved, but there also must be a willingness to compromise, to seek some middle ground, so the outcome becomes a win-win situation for both sides. While there are humane rights that need to be protected, there are oftentimes human rights that need to be considered. The absence of seeking understanding and moral clarity in any purposeful defense eventually wrongs both parties and perpetuates an unresolvable, adversarial, oppressor-oppressed battlefield. Voters can flock to the polls, protestors can chain themselves to logging trucks, court orders can be obtained, and petitions can be signed and filed, but unless a place is defended in a way that ultimately gets both sides to give a little and get a little, the need to win will far outweigh the bigger issue—*doing what is right and doing it right now.*

The release of wolves into Yellowstone National Park provides a prime example of how to successfully address both sides of a volatile issue, although the project is still in its infancy and continues to incite strong feelings on both sides. Right from the start, however, the wolf-release program sought humane as well as human justice and continues to uphold a compromise that values the views of all concerned. Working in tandem, the National Wildlife Federation and the Defenders of Wildlife have promised that any rancher who loses livestock to a reintroduced wolf can receive financial compensation for the cost of the lost livestock. True to their word, calls from ranchers have been promptly investigated and checks have sometimes been cut on the spot. The organizations have promised financial compensation for any landholder who allows a wolf to den on his or her property as well as vows to destroy any "nuisance wolves" that cannot be successfully dealt with. In return for such human considerations, the organizations have also promised that anyone who shoots a reintroduced wolf would be subject to prosecution; after reintroduced "wolf 10" was shot illegally, the gunman was convicted under provisions of the Endangered Species Act, which included a fine as well as a six-month imprisonment. Because the objective of the wolf reintroduction program was always based on education about the wolves as well as on providing an open forum in which to understand and address the concerns and issues of those who were opposed to the program, both parties have, thus far, been able to effectively defend their positions.

Perhaps the greatest defender of a place was Chief Seattle of the Duwamish tribe; most people are familiar with Chief Seattle because words that have been attributed to him have served as the basis for the environmental movement as well as have appeared in countless eco-pamphlets, calendars, environmental fund-raising letters, children's books, and books of quotations. "How can you buy or sell the sky, the warmth of the land?" asks Chief Seattle. This question was taken from a speech the Chief is said to have delivered in 1854 to the newly arrived Commissioner of Indian Affairs for the Washington Territory. "The idea is strange to us," Chief Seattle tells the commissioner. "If we do not own the freshness of the air and the sparkle of the water, how can you buy them?" While the authors of *The Native Americans: An Illustrated History* note that in the spring of 1992 the *New York Times* ran a story that determined that the Chief Seattle speech was a fraud and had actually been produced by a Texas literature professor in the early 1970s, the eloquent sentiments expressed in the Seattle speech have, for decades, accurately and eloquently symbolized the dilemma between how best to defend the land and all of its living creatures while living and surviving on the land:

> *Every part of this earth is sacred to my people. Every shining pine needle, every sandy shore, every mist in the dark woods, every clearing and humming insect is holy in the memory and experience of my people. The sap which courses through the trees carries the memories of the red man. . . . We are part of this earth, and it is part of us. The perfumed flowers are our sisters; the deer, the horse, the great eagle—these are our brothers. The rocky crests, the juices*

of the meadows, the body heat of the pony, and man—all belong in the same family. . . . If we sell you our land, you must remember, and teach your children, that the rivers are our brothers and yours, and you must henceforth give the rivers the kindness you would give any brother.

. . . We know that the white man does not understand our ways. One portion of the land is the same to him as the next, for he is a stranger who comes in the night and takes from the land whatever he needs. The earth is not his brother, but his enemy, and when he has conquered it, he moves on. He leaves his fathers' graves behind, and he does not care. He kidnaps the earth from his children. He does not care. His fathers' graves and his children's birthright are forgotten. He treats his mother the earth and his brother the sky as things to be bought and plundered, sold like sheep or bright beads. His appetite will devour the earth and leave behind only a desert.

The expedition that journeyed into the deep wilderness of the Yellowstone Plateau in 1870 experienced a similar environmental dilemma. The party of adventurers, led by thirty-eight-year-old former Montana territorial governor Nathaniel Pitt Langford, invaded a wilderness never before seen by human eyes. They climbed mountains, hiked to the base of waterfalls, and stared open-mouthed at the spectacle of a sky-rocketing stream of hot water they named "Old Faithful." One night, as the party sat around their campfire smoking cigars, they chatted about how they might best divvy up such a tremendous piece of the earth. They were sure that the real estate would "eventually become a source of great profit to the owners." But one member of the party, Cornelius Hedges, shook his head in disagreement. "Mr. Hedges said there ought to be no private ownership of any portion of that region," Langford later wrote in his journal, "but [that] the whole ought to be set apart as a great National Park, and that each of us ought to make an effort to have this accomplished." Even though Langford had made the trip to Yellowstone as a scout for the Northern Pacific Railroad, which had wanted new territory in the vast empire of the western lands in which to haul visitors for a price, and even though Langford viewed Yellowstone as a potentially profitable tourist attraction for his employer, he was one of those who pushed hardest to have the Great Geyser Basin set aside as a federal reserve. On March 1, 1872, President Ulysses S. Grant signed the bill that made Yellowstone the first national park. "However mercenary the reasons for its creation," writes environmental journalist Jim Robbins in *Last Refuge: The Environmental Showdown in Yellowstone and the American West*, "America's first natural reserve soon became a worldwide example of the selfless preservation of natural wonders, of far-sighted social policy, and a gift to humanity."

Today there is a growing strength in environmentalism, animal rights activism, conservation education, and historic preservation movements that are seeking to preserve and protect not only the land but its inhabitants—man and beast—in ways that thrive on awareness and understanding rather than on enforcement and self-serving motives. Sometimes this happens when a large group of people bands together for a common purpose. Native Americans, for example, have organized themselves into a multitude of small groups—with names like Native Americans for a Clean Environment, Diné CARE (Citizens Against Ruining Our Environment), Anishinable Niijii, and the Gwich'in Steering

Committee—to face down huge waste dumps, multinational mining and lumber companies, and the U.S. Office of the Nuclear Waste Negotiator. Too, Native Americans have banded together as a culture to press for the return of the remains of Indian ancestors for appropriate burial on their homelands, such as the release of the bones of two Connecticut Pequot Indians, which were unearthed in 1918 during digging for the U.S. Naval Submarine Base in New London and have been lying in state in the Smithsonian's repository of American Indian bones. Lakota Sioux in the Southwest have sought to restrict recreational climbing on their culturally sacred Devil's Tower, also the country's first national monument, during the month of June, a time that is sacred to theirs and many other tribes.

Sometimes defending a place attracts national attention, such as when musician Don Henley came to Concord, Massachusetts, with the idea of raising money to save the woods surrounding Walden Pond. Saving the Walden Woods—a project that eventually accumulated ninety-six acres in Concord, Massachusetts, many of which were wrestled out of the grasp of developers—pitted preservationist against developer as well as Thoreauvians against residents who were already weary of the notion of yet another preservation movement going on in a town that was already crammed full of historically preserved sites: Walden Pond and the surrounding woods, where Thoreau constructed his crude cabin; Battle Road, the route of the colonial minutemen who eventually fought the British at the Old North Bridge, site of the first bloodletting of the American Revolution; Minuteman National Park; and the homes of Ralph Waldo Emerson, Nathanial Hawthorne, and the Alcotts, along with numerous museums and a graveyard stocked with the remains of literary geniuses. Henley and his campaign were able to successfully raise the $15 million necessary to preserve the site.

But, more often than not, defending a place happens on a much smaller and less publicized scale, when a group of concerned citizens simply bands together for a common purpose and focuses their time, attention, and energy on something that has a great meaning in their lives. The residents of Big Bear Lake, California, for example, built a tiny zoo and raised money through an unusual campaign that started with bake sales, aluminum can drives, and storefront raffles in order to save the lives of a mother grizzly and her cubs, who were slated to be destroyed by Montana wildlife officials because they were "nuisance bears"— caught repeatedly raiding garbage cans and dude ranch cabins on national forest land just outside Yellowstone Park. Volunteers designed and built a 10,000-square-foot, pine-shaded grizzly habitat at the community-run Moonridge Animal Park after learning of the bears' plight; they decided it would be fitting for the community to save the lives of a few of its namesakes. In another example, a small group of devoted skiers put their cash on the line to rescue tiny Mad River Glen in Waitsfield, Vermont, from an uncertain fate and have since worked together to keep the mountain open as a ski area. The co-op elected to refrain from widening and grooming their trails as bigger ski areas do and to stick with Mad River Glen's 1950s premise that skiing should harmonize with nature, not dominate it. Only a bare minimum of man-made snow is used to cover the base so skiers can get to the lifts and then ski down trails that twist and turn along the mountain's natural contours. And, in yet another example, David Carroll, who was originally trained as a painter, has studied turtles so extensively over the years that he is now recognized as one of the leading experts on the New England wetlands where turtles live. He raised ten thousand dollars from the sales of his first book, *The Year of the Turtle,* for the World Conservation Union's global turtle-recovery project and has become one of the most important voices for conservation of wetlands that not only shelter his beloved turtles but also ensure clean drinking water and protect humans against floods. It was his testimony that led to federal protection of a twenty-three-mile stretch of the Lamprey River in New Hampshire and state protection of four species of turtles; in addition, his surveys stalled development of a $19-million mall as well as a fourteen-mile, federally supported highway bypass that would have destroyed vital swampland.

One man or one small community can defend a place as effectively as a huge, internationally recognized organization can. All it takes is desire, determination, and the belief that there is a higher purpose to this life on Earth: that you are not just here to enjoy life while you can, but you are also here to help meet the needs of all living things, to enhance the quality of life for all living things, and to protect the future of those who will follow you by living your life as a steward to the whole Earth, not just the parts that happen to be useful to you. As His Holiness Tanzin Gyatso, the fourteenth Dalai Lama of Tibet, has written, "It is not difficult to forgive destruction in the past which resulted from ignorance. Today, however, we have access to more information, and it is essential that we reexamine ethically what we have inherited, what we are responsible for, and what we will pass on to coming generations. . . . We have the capability, and the responsibility. We must act before it is too late."

Some Ways to Seek Natural Preservation

1. Support school programs that encourage students to create and contribute to "living classrooms," such as gardens and forest paths, as well as teach humane care of creatures held in captivity. Learn what your children are learning; make sure all aspects of nature and wildlife are presented by teachers without bias.

2. Maintain the health of trees on your property as well as work with town officials in ensuring that town trees are tended to in a timely fashion.

3. Organize a community clean-up group to keep abandoned urban areas as well as forests, parks, and other suburban and rural areas free from litter and other debris. Work with town officials in converting small, unsaleable tracts of land into public gardens or parks. Share weeding, watering, mowing, and planting chores with others.

4. Join a watershed association or start one for a river in your area so you can help with water-quality monitoring, participate in river cleanups, and add your voice in creating strategies to maintain a river's health. Nominate your favorite river for the federal Wild and Scenic Rivers program, which keeps rivers free-flowing and prevents water diversion; the program requires that local groups work with municipalities in protecting designated rivers and adjacent land so communities become river stewards. For more information, contact the National Park Service, which oversees the program.

5. Become familiar with hiking trails, both public-accessed as well as those that are privately accessed. Leave no trace of your visit; carry out what you carry in, including bodily waste. Report any trail abuse—littering, setting fires, illegal hunting, and so on—as well as post warning signs if cars have been broken into or you or others have been accosted by strangers. Be sure

mountain bikes and cross-country skiing are allowed on the land prior to your use; stay on appropriately marked trails. Keep your dog on a leash at all times, and pick up after your pooch; respect "No Dogs Allowed" signs and leave your dog at home, no matter how well-behaved your dog is.

6. Elect public officials who share your sentiments about the land and its living things. Appointed town and city officials should be closely monitored, particularly when they serve on boards that determine how natural resources in the community are used. Make sure the composition of boards such as the Fish and Wildlife Commission is made up of hunters and anglers as well as naturalists and conservationists.

7. Know what is going on in your community. If you don't know, ask. Never assume that someone else is taking care of things in the best way possible. Get active. Find out, for example, where your trash goes after pick-up as well

as how your community contributes to its recycling program.

8. Help someone understand a different side of an issue. Rather than attempt to sway an opinion or change a sentiment, seek instead to raise awareness. Confrontation or shock-and-shaming techniques ought never be used; being offensive will not help your defense of anything. Support those organizations that defend a place or a living thing through the dissemination of factual and educational information.

9. Read labels. Find out what some of the ingredients are that you cannot even pronounce. Consider whether you want to continue to eat products that contain such ingredients. Whenever possible, support organic farmers and their products.

10. Report any abuse of an animal to the appropriate authorities immediately. Defend every living creature's right to humane treatment and care.

This Wasn't a Part of the Plan:
Supporting Unanticipated Natural Effects

"Before the beach became a fertile habitat for countless birds, clams, crabs, insects, fish, worms, amphibians, reptiles, and more (and before it became a place for mammals including man to visit) it had first to become—well, a beach. The geological processes that made it began eons ago and, abetted by evolutionary botany, show no signs of stopping. This place, ancient and perpetual, continues to change. Enchanted, man has tried to arrest the beach, to settle here and alter it for his convenience. Yet it continues to grow and change to suit itself. It always will."
—science and humanities writer and reporter Philip Kopper, from *The Wild Edge*

Medical studies say that humans should not live beneath the thousands of miles of power lines that crosshatch the landscape; it is not healthy. But humans need electricity, so they build the power lines and then step away from them. Viewed from a distance, the skeletal towers appear forbidding—stark, cold, eerie—towers of doom that rise high above the nearby forest. The power companies visit frequently to wage their perpetual war against the growth of trees beneath their precious lines; they mechanically prune or use herbicides and then leave, not wishing to linger too long in the forbidden area. In the silence that is restored after their departure, the incessant crackle above the power lines can be heard, proof that the electricity is surely snapping its way from source to terminus.

But something else happens after the power company workers leave, something that few except naturalists, environmentalists, birders, hikers, bikers, and hunters know about and have witnessed for themselves. The land beneath the power lines springs to life—glorious life—in the wonderful, wildlife-sustaining habitat that has been created and is being maintained by man. Known as an "edge habitat"—a place where the open land meets the nearby "climax" forest, which is full of trees and therefore maintains a dark and barren floor—the open land beneath the human-unfriendly power lines proves to be extremely wildlife friendly. Since wildlife depends upon open space for hunting, foraging, and reproducing, low-growth areas such as grasslands, shrubs, and meadows play a vital role in supporting all sorts of wildlife. While man's thoughtlessness and greed over the centuries has snatched much valuable open space away from nature in order to create human-friendly developments, the land beneath the power lines has, unknowingly, returned a valuable resource to natural living things and, in so doing, has repaid nature a tremendous debt.

"Most of the wildlife I've seen in New Hampshire, I've seen beneath the power lines," says Ellen O'Donnell, a wildlife biologist and head of the environmental department for the Public Service Co. of New Hampshire. "Deer, fox, grouse, warblers. It's rich, open land." John Lanier, of New Hampshire's Department of Fish and Game, echoes O'Donnell's assessment: "You've got the small animals that live there and you've got the bigger ones that come out of the forest to graze." Birdwatchers know that the space beneath the power lines is one of the few places in which to spot the prairie warbler; as well, the space is also prime habitat for yellow warblers, rufuos-sided

wild rose, Provincetown, MA, 1956 © David Vestal 1997

towhees, eastern meadowlarks, red-wing blackbirds, indigo buntings, and bobolinks. Bears come for the berries; it is fairly common to spot a mother bear and her cubs munching their way through the bountiful natural farmstand that can feed birds and animals for months—bushes droop low with warm, juicy, sun-ripened elderberries, strawberries, blackberries, checker berries, black cherries, blueberries, raspberries, and wild-pasture roses rich with rosehips. Moose and deer graze on tender shoots and grasses; their first feed of the spring is often provided by the land beneath the power lines because the exposed, low-growth area melts out faster in the spring than in the nearby dark forests, which block out shoot-encouraging sunlight. Great concentrations of butterflies thrive beneath the power lines, including the rare Karner Blue butterfly, as well as big silk moths such as the Luna, the Promethea, and the largest in North America, the Cecropia Moth. Small mammals such as shrews and rabbits happily multiply in the constantly buzzing community, making it a hunting heaven for their predators—fox, coyote, and red-tailed, red-shouldered, and broad-winged hawks. With a little help from man, each species can easily maintain the natural balance of the edge habitat, giving a boost to the wildlife cycle. Because of this, the power industries are becoming more and more diligent at not just keeping trees at bay but also protecting the berries, bushes, and grasses; many have launched studies of a variety of herbicides in order to find the ones that will work best. Too, environmental groups, hunters, birders, and nature lovers are interfacing more with the power industry, first by defending man-made power lines against human encroachment and then by working out creative solutions that connect rather than disconnect man from nature.

More often than not, if anything goes wrong in or with nature, it is because of man. Even some of the most well-intended conservation efforts have backfired; witness the overpopulation of Canada geese that now infest golf courses as well as bring traffic to a grinding halt on major highways as multi-generational families of geese as well as their large, extended family communities slowly waddle across several lanes of traffic in order to sample the tasty offerings at the Median Strip Café. Too, this nation's twelve million deer owe a debt of gratitude to human beings who, in their misguided wisdom, determined that the coyote and the wolf needed to be eradicated; with such predators out of the picture, the deer population has become so massive that deer now invade urban and suburban turf, munching their way through costly landscaping and backyard gardens, or they slowly starve to death in the long, cold winter months. The deer are joined in cities and suburbs by raccoons, skunks, and opossum,

who have been enticed into such "prime" habitats that offer them yards, parks, and golf courses (more man-made, nature-friendly "edge habitats"), an abundance of garbage, and a lack of predators. One suburban couple became so tired of having wildlife raise their families in their basement that they began to block all entrances to the basement with heavy stones. "We'd use heavy stones in front of the holes," explains the wife. "But the next morning we'd find them moved, as if some animal miracle had occurred. We love animals, but we just wish they'd stop raising families inside our house."

At Walden Pond, Thoreau used to fling open the door to his crude cabin and pull up all the windows not only to let in the light and fresh air, but also the birds and squirrels. Thoreau would not even begrudge a wasp access to his home; today a single wasp buzzing in a family's den is like a live grenade that has been tossed into the home, capable of shocking a family into a fit of frantic activity. American nature writer John Hay once observed, "I cannot forever keep out the woodpecker that mistakes my house for a dead tree. For that matter, why should I object to a flicker banging away on the roof when what it is doing is proclaiming the triumph of spring?" Whereas you may feel that you are rapidly losing the battle against things that want to come into your home and your yard—squirrels, wasps, house sparrows, spiders, field mice, and countless other living things—what is happening in reality is that you are being presented with one of nature's laws, a law that mankind has refused to acknowledge over the years, from the time when man first tried to fit in with nature, then to get along with nature, and then, finally, to control nature. This natural law states: *Because nature abhors a vacuum, nature will always fill whatever vacuum has been created according to nature's desires.* As is most often the case, the vacuum is filled with what is usually most

unwanted and least likely to be effectively managed by mankind. A prime example of this is the so-called "miracle vine," which was once promoted as a soil protector in the Southeast and is now an unstoppable weed. Historian Edward Tenner, who details countless technological and nature-related snafus in his book *Why Things Bite Back: Technology and the Revenge of Unintended Consequences,* describes this weed as now "...able to overwhelm almost any stationary object — unmoved automobiles, sidetracked railroad cars...even (so says Southern folklore) unconscious drunks."

This wasn't a part of the plan is the most oft-heard defensive response given when human interactions with nature—even those made with the best of intentions—go awry. In the case of human technology versus nature, more often than not even though nature is greatly impacted at the outset, sometimes irreversibly and sometimes reversibly, nature ultimately wins. Witness the case, for example, of the South American fire ant. Unknowingly brought into the southern United States as shipboard stowaways in the 1930s, the fierce biting ants, which like to sink their mandibles into people and livestock as well as electrical insulation—they sometimes even knock out traffic lights with a simple bite—were slated for eradication following the development of DDT and other superpesticides. After three decades of spraying fire-ant territory with the killer compounds, the United States government begrudgingly admitted defeat. The pesticides that were being used to eliminate the maliciously destructive insects instead eliminated their natural enemies, leaving the pests more virulent than before and creating an even a bigger problem. To this day, fire ants are still uncontrolled.

Countless other examples abound. A nineteenth-century French naturalist who brought gypsy moths to the United States in the hopes of founding a silk industry unknowingly allowed some of the caterpillars to escape; their descendants persistently ravage forests across North America. The Glen Canyon Dam, built in 1963 to hold back Lake Mead, drastically reduced the size of the Colorado River as well as disrupted the valuable ecosystems and recreational activities it once supported. As a result, in 1996 the Bureau of Reclamation authorized $4.5 million to flood the canyon in what people hope will be the first of many supervised releases that will help to restore riparian habitats to predam conditions, replenish rapidly disappearing sand beaches, drive off nonnative fish species, and promote the proliferation of dozens of native species in a technological mimicry of nature's own springtime floods. Not only are environmentalists and utility companies pleased with the plan, but also whitewater rafters, who, in the last few decades, have had to grow accustomed to a kinder, gentler stream. "You won't want to be out there if you don't know what you're doing," comments Andre Potochnik, a Colorado river guide. "We're finally going to see the real Colorado River—not a shy imitation."

The biggest problem with the strong human drive to perfect the presumed imperfect world in which they live is that humans are often imperfect themselves at figuring out the repercussions of what they intend to do. Virtually everything that mankind has done has had its unintended consequences. Engineers have built breakwaters, seawalls, and groins to keep surf from eroding beaches at popular resorts, but the man-made structures have interrupted wave-driven drift and stolen sand from downshore beaches. The most tragic consequence of such engineering activity was the 1970 flood in Bangladesh that claimed more than 225,000 lives; in large part, the breakwaters and other flood-control systems that had been designed to protect human life from the ravages of nature had actually encouraged widespread settlement of flood-prone land. In the United States, the long-held strategy of fire suppression, symbolized by the fifty-year-old Smokey the Bear campaign, has actually had an incredibly detrimental effect. Firefighting techniques have become so effective today that small forest fires are usually doused long before the debris that eventually fuels large, unmanageable ones can be consumed. After the devastating New Mexican "Hondo" fire of 1996, which consumed roughly 7,500 acres, raced faster than any human being has ever run, tore a miles-wide path through Carson National Forest, and created a wall of flame a hundred feet in the air, analysts pointed out that decades of relentless firefighting have had an unnatural effect on forests that actually need occasional fires in which to rejuvenate and grow. "A bad fire year is a year when there aren't a lot of fires," jokes Brian Morris, a district ranger from Six Rivers National Forest in northern California. While suppressing fires has become a tradition in America's West and firefighters are beloved figures, praised for their respect for the land and the efforts to which they go to demonstrate it, in academia they have become symbols of a failed fire-suppression policy and the cause of the ferociousness of today's forest fires. "Fire is good because it recycles nutrients," explains Malcolm Zwolinsky, a professor of renewable natural resources at the University of Arizona at Tucson. "It maintains certain types of ecological systems. When fires occur, they are surface fires. They remove surface materials and do not get up into the crowns. When you change the period of fires, from maybe seven to ten to fifty years, you have imposed a substantial change on the ecological system. When fire occurs, it is going to be much more damaging." The consensus opinion today is that prescribed fires should be periodically set by trained firefighters in order to clear out ancient kindling that could eventually propel a wildfire into a devastating, uncontrollable blaze.

These few examples reveal that mankind has not only consistently ignored the natural law of nature's need to fill vacuums, but also the human law, first espoused by Army Captain Edward Murphy, Jr. in the 1950s that "If there's more than one way to do a job and one of those ways will end in disaster, then somebody will do it that way." Or, as Murphy's supervisor, Major John Paul Stapp, later paraphrased in a news conference, "If anything can go wrong, it will." The principle has stuck, along with credit to its originator, and has since come to be known as Murphy's Law. While Murphy's Law has, over the years, been interpreted as a pessimist's view of the consequences of one's actions, it is, in reality, just a simple reminder—as is nature's need to fill any vacuum—that greater efforts need to be made in designing things, in creating things, in controlling things, in changing things, and in attempting to resolve problems created by things so that unanticipated error, if not always impossible to prevent, would at least be less likely or not particularly harmful, with the best-case scenario being that in which the consequences of any action would not only be considered and, therefore, anticipated, but also accounted for in the process of designing any solution.

Rarely is the resolution to any situation that involves the natural world ever foolproof or permanent. Mankind has learned time and time again that nature is unpredictable, adaptable, fragile, determined, interdependent, resilient, vulnerable; as a result, there are simply too many unknowns in the equation *nature divided by man*

equals x to assume that there is only one answer, one outcome, one solution. Most solutions, as can be seen throughout the history of mankind's interactions with nature, have had a negative impact and often require going back to the drawing board (and bank accounts, board rooms, and courts of law) to put more thought, planning, time, energy, and finances into rethinking a solution that needs to deal not only with the original problem but, as well, the newer, more complicated problem that has resulted from the attempt to resolve the problem in the first place.

Some solutions are thankfully reversible—such as the case with the restoration of the Colorado River; some are resolvable—such as setting prescribed fires to alleviate potentially destructive wildfires; and some, in rare cases, bring about such unexpected and pleasantly unanticipated boons that mankind can learn valuable lessons from them about how to create and strengthen the relationship between the human world and the natural world so both worlds can benefit.

The Rocky Mountain Arsenal, just northeast of Denver, is fast becoming a "poster-child" of toxic recovery for man and nature. During World War II, tons of mustard gas, chlorine gas, and other chemical munitions were produced in the arsenal for possible use against the Nazis. Deadly incendiary bombs were packaged for bombing raids on Tokyo. After the war, private chemical companies produced tons of now-banned agricultural pesticides there. And, during the time in which Americans were not in conflict but were constantly reminded of its potential during the Cold War—a terrifying time in American history when the periodic test tones of the Emergency Broadcast System made radio listeners' pulses race, when grade school civil defense drills were practiced with greater frequency than fire drills, and when backyard bomb shelters were built and kept well-stocked with fresh batteries, water, and canned goods—the U.S. Army arsenal produced massive quantities of nerve gas.

The Rocky Mountain Arsenal became not just an eyesore or a stinking, groundwater-contaminating neighbor of Commerce City, but an enormous dump of toxic waste—what is today identified as a Superfund waste cleanup site with a two billion dollar price tag that will keep workers in protective clothing and technicians attempting to decontaminate the billions of gallons of groundwater a year employed long into the next century. And yet, within eyesight of the cleanup crews are bass fishermen angling in a pond and school children on a field trip hoping to catch their first glimpse of America's symbol, the bald eagle. For, at the same time that the United States government is responsibly cleaning up the mess it made, it is going one step further by striving to make the arsenal an amenity to the area; in

tern, Provincetown, MA, 1957 © David Vestal 1997

fact, the United States Fish and Wildlife Service is transforming the arsenal into a wildlife refuge in what many believe will be perhaps the premier urban wildlife refuge in the nation. Tom Kenworthy, a reporter for the *Washington Post,* notes that "Residents of the Denver metropolitan area who for years feared the arsenal as a health threat are now beginning to see it in another light: as the largest chunk of open space in an increasingly crowded urban landscape, as a place to watch eagles in their winter roost, even as a place to catch trophy-size bass." Members of the Denver chapter of the National Audubon Society marvel at the unusual species they have observed at the refuge, including ferruginous hawks, currently candidates for the endangered species list. Acres of short-grass prairie in the twenty-seven-square-mile plot of land (of which only about fifteen percent is directly affected by the byproducts of weapon and pesticide production) provide a cozy home to almost three hundred different species of wildlife. Catch-and-release ponds are being kept well-stocked with bass. And the extensive network of biking and hiking trails, which steer clear of the contamined core of the Superfund area, allow access into a unique wildlife oasis that flourishes in an area many thought to be hopelessly barren and permanently impacted.

A similar back-to-nature transformation occurred in 1996 on Noman's Land Island—a prime piece of ocean-surrounded property just three miles north of Martha's Vineyard, Massachusetts—when, after an extensive summer cleanup, the island was turned over to the federal agency known as The Great Meadows National Wildlife Refuge. The island, with its 628 acres of ponds and brush, has, since 1942, served as a practice bombing and strafing target for United States Navy pilots, living up to its name of Noman's Land after forty years of being strafed by up to five attacks per week. Were the island in "mint condition" and not littered with unexploded bombs, sharp and rusted metal fragments from rockets, "paint" bomb casings, and the scattered remains of targets such as aircraft fuselages, it would be worth nearly $500 million to developers. But the government—in an attempt to rectify the havoc and disruption it has caused not just to an ecosystem but also to the residents of Martha's Vineyard (where people pay up to two million dollars to live a peaceful, oceanfront island existence); to those who live on Cuttyhunk Island, twelve miles to the northwest, and to those who earn their living from fishing who have often been chased away from the area by Phantom jets—is abiding by the island's 1666 colonial charter that determined the island belonged to "no man." The island will remain off-limits to the public as it has for decades and, instead, become the exclusive home to peregrine falcons, bald eagles, and a host of migratory shore birds.

These two instances of natural transformation provide good examples of positive, life-sustaining outcomes that were never part of an original plan; few could have ever predicted at any time that such areas would ever become what they are today. Such outcomes have clearly been carefully examined, explored, and then enacted after an original solution proved ineffective or, in the long run, unsuitable for mankind. But what about those unanticipated outcomes that have produced surprising results—results that have unknowingly changed a human attitude about the relationship between nature and civilization in positive and mutually beneficial ways—outcomes which, like the nature-friendly power line habitat created by utility companies, were never part of anybody's original plan?

Take, for example, what happened after the Congress Avenue bridge in downtown Austin, Texas, was rebuilt in 1980. The new concrete slab had expansion joints—

slots running lengthwise along the underside—which engineers did not realize would provide perfect temperature and humidity conditions in which bats could set up home. Since bats are quite content when they are crowded together—up to two hundred bats roosting in a square foot is not uncommon—the colony under the bridge rapidly grew. Soon, an estimated 1.5 million bats were happily residing under the bridge. At first city workers and residents were alarmed at the "invasion" of bats; there were heated discussions about how to get rid of them. But then Bat Conservation International, a group founded in 1982 by bat advocate and wildlife conservationist Merlin Tuttle, was called in to help change public opinion. Tuttle calmed the people, first dispelling the common myths that the bats would fly into people's hair or bite their necks. He presented residents with the reassuring fact that fewer than 0.5 percent of bats carry rabies and that virtually none of them ever comes in direct contact with humans. But when he conveyed the figures on the quantity of insects bats consume in their nightly feeding, Austin citizens quickly embraced the organic, pest-controlling, bridge-inhabiting colony; soon, they were proudly bragging to anyone who would listen that *their bats* could consume *between fifteen and thirty thousand pounds of insects*—moths, beetles, and flying ants—*every night.* Crowds of bat-loving citizens—sometimes even entire families—began to ritualistically gather together in the spring and summer evenings, just before sundown, near the bridge or spread out blankets on a knoll above a nearby lake to wait for "the big show," when the Mexican free-tailed bats would emerge from their roost beneath the bridge and then head out for their "night on the town." As the sun would begin to sink behind the hills, the crowd would grow hushed; in the dusky light, they would then watch in awe as the rush of millions of bats wheeled out into the night. Over the din of flapping wings and bat chatter, people would shout and cheer; the incredible "fly-out," which lasts for half an hour, continues to draw adults and children together to witness the incredible spectacle of how man and nature can live together.

For nearly a century mankind has considered itself to be the manager of the biosphere. Mankind has caused continental icecaps to recede, wildernesses to vanish, prairies to turn to dust and blow away, oceans to rise, forests to vanish, wetlands to dry up, rivers to flow in different directions, the ozone layer to be depleted, and species to disappear—as well as destroyed the land and the heritage of many of the world's native peoples. Mankind has threatened biology and evolution, polluted and used the land and its living things as political weapons, and employed technology for destructive purposes. For a time mankind believed

that, despite such things, it was a good manager, for its own species continued to be fruitful and multiply. But then mankind began to see how ineffective a manager it truly was—how inept, partial, and destructive it was—when its own life support began to be threatened, when it finally saw that its actions had not only had a detrimental impact upon nature but also upon itself. Mankind had unknowingly started to poison itself; humanity was beginning to weaken and die. Birth defects, cancers, tumors, and genetic disorders signaled the end of living on the planet and the beginning of the need to survive.

Today mankind has finally begun to realize this, that it needs to resign as manager of the universe and, instead, develop a sensory, perceptual, and psychological approach to living that grants all living things not just the right to live on the planet but also to live in it or within it. Mankind now knows that it has to become a conserver and preserver not only of its own future but also of the future of the life around it.

Mankind is slowly learning that for every action it takes, there is a reaction. Too, mankind is also realizing that to get the most beneficial and least harmful reaction, it must use more responsible decision-making long before it takes action and then, through whatever actions it takes, benefit itself as well as provide better management of the biosphere. Mankind knows now that it must create a new vision not only of itself, of all of humanity, but also of nature, and then must ascertain what role each plays in each action and in each reaction and *defend these roles*.

It is simply no longer acceptable for mankind to merely shrug and mumble, "This wasn't a part of the plan" when confronted by damage, destruction, or extinction it has caused. Nor is it acceptable for mankind to beam with pride at beneficial but unanticipated outcomes to its actions. Mankind needs to see itself as an equal participant in life on the planet, as integral and yet as expendable as even the lowliest weed that it would seek to eliminate. Mankind needs to accept that it is merely one link in the endless chain of life. Mankind must learn how to live not just in greater harmony with nature but also in the greatest harmony of all, in a deeply symbiotic relationship with nature that reflects the essence of Chinese sage Chuang Tzu's observation: "I do not know whether I was then a man dreaming I was a butterfly, or whether I am now a butterfly dreaming I am a man." What this means is that a power company can be a naturalist, a defense contractor an environmentalist, a hunter a defender of wildlife, a contractor a conservationist. Mankind can assist nature and nature can assist mankind; nature and mankind need to live together in the planet in a life-giving, life-sustaining partnership that *is always* a part of the plan.

Some Ways to Support Unanticipated Natural Effects

1. Use natural methods to make your environment more comfortable. For example, for the past twenty years the Chamber of Commerce in Wells, Maine, has held an annual spring sale of dragonfly nymphs, which are known to have a voracious appetite for mosquito larvae. Rye, New Hampshire, breeds mummichogs—another mosquito-eating critter—and then releases them by the thousands into the prime marshland that surrounds the community. Volunteers in many New England communities build and maintain houses specifically designed for bats, swallows, purple martins, and wrens, which eat the bothersome pests. Citronella candles work just as well as mosquito repellents. Even though pests will always be a part of nature, they can also always be dealt with in more natural ways.

2. Recognize that there is a useful purpose to every living thing. If you are not aware of this purpose, find out what it is. The dandelion, for example, which you may consider to be a weed and thus seek to eliminate from your lawn, has fed and healed humans for thousands of years. Its leaves make an excellent salad, equal to endive. Infusions of dandelion were used in monasteries and by village herbalists for kidney and liver problems, skin diseases, and stiff joints. Its milky juice was rubbed on pimples and warts. And, of course, there is the infamous concoction of dandelion wine, which has warmed people for centuries.

3. Remember that everything in nature acts in conformity with law, even though nature's laws number very few. Nature does nothing uselessly, for instance. Nature never overlooks a mistake nor makes the smallest allowance for ignorance. Nature does not proceed by leaps. Nature abhors uniformity. Nature offers neither rewards nor punishments, but consequences. Nature is monotonous in its orbits,

tides, sunrises, and sunsets. And nature's rules have no exceptions. *Respect the laws of nature.*

4. Support the construction of biking and hiking trails that make use of abandoned railroad beds or other deserted properties that nature has reclaimed. Work with abutting landowners to ease their fears over disruption to their lives, destruction to their property, or invasion of their privacy. Create a plan that anticipates and then handles the needs of everyone who might be impacted.

5. Refuse to curse the wind, rainy days, snowstorms, leaves that pile up in your back yard, the lack of rain, high humidity, an early frost, and countless other bits of weather or natural circumstances over which you have no control. There is a purpose to each and every natural action. When Henry David Thoreau was at his retreat on Walden Pond, he wrote of a life lesson he learned after enduring a stretch of rainy spring days. At first, as he watched the

rain, he expressed his dissatisfaction that he was housebound, but then he voiced his happiness at seeing the bean seeds he had recently planted being watered. Then, as the rain continued, he pondered the survival of the seeds. After much thinking, he observed that even if the seeds rotted in the ground, the rain, "...will still be good for the grass on the uplands, and being good for the grass, it would be good for me."

6. Rather than plant a neat, orderly flower garden, purchase a variety of seed packets and then mix all the seeds together. Set aside an area of your yard in which you haphazardly plant the seeds. Then wait and see what grows. Or make a pile of rocks in another area of your yard. Watch what sets up home in the rocks. In the summer, turn over a few rocks and observe what lives under the rocks.

7. Consider wildlife to be a natural attraction, no matter where you live. Remember that deer, fox, porcupines, woodchucks,

skunks, beaver, squirrel, opossum, rabbits, and raccoons arrived long before you moved in. Recognize who is the real interloper and then strive to work out a harmonious relationship with those living things that share your property with you. If you must remove a critter, do so by using a catch-and-release trap. Even a skunk that is captured in a well-designed trap cannot spray while inside the trap so it can be safely relocated.

8. Accept the landscape as a moral and spiritual place. Cultivate it but do not try to remake it. Aesthetic pleasure ought never to dominate nature. If you must take down a tree, pull up a shrub, or mow down meadow grass, plant another tree in a different location, relocate the shrub, and allow meadow grass to grow freely in another area. Each are homes to many natural things that are vital to your life.

9. In many locations, local bees are becoming scarce in supply. The varroa mite, which began killing honey bees around the country ten years ago, was spread largely by mail-order bees sent to beekeepers; too, long winters have a devastating effect on bees. To sustain fruit orchards, tens of thousands of bees are needed for all the pollination. Whenever possible, adopt an attitude of "live and let live" with honey bees on your property unless you are allergic to them, they are endangering your family, or they are boring their way into the walls of your home. If you live near a field or other open area, set up a hive that will not only provide shelter for honey bees, but also facilitate the pollination of flowering plants and trees near your home.

10. Never allow a town, city, state, or federal agency to give up on any area that has been negatively impacted from past decisions. Work together with your local government or join a concerned group that can help to restore the natural balance in such areas, even those that have been quite severely impacted. Remember, amidst decay and ruin there can always be growth.

Wildlife with an Attitude:
Embracing Nature's Outcasts

"We reached the old wolf in time to watch a fierce green fire dying in her eyes. I realized then, and have known ever since, that there was something new to me in those eyes—something known only to her and to the mountain. I was young then, and full of trigger-itch; I thought that because fewer wolves meant more deer, that no wolves would mean hunters' paradise. But after seeing the green fire die, I sensed that neither the wolf nor the mountain agreed with such a view"
—naturalist and writer Aldo Leopold, from *A Sand County Almanac*

"Standing alone, I thought, as I had so often before, of Aldo Leopold and 'the fierce green fire' he had seen in the eyes of a dying wolf. That fire had been extinguished over and over,
in a reign of arrogance and misunderstanding that led to the extermination of wolves from nearly every state. But sixty years after Yellowstone's last wolf pack was slaughtered, we were bringing the species back, empowered by the provisions of the Endangered Species Act.... In returning wolves to Yellowstone, we made the natural system whole again. We said we could not, and should not, control the land only for ourselves. We said that the howl of a wolf had a value at least as great as a slab of beef or a rack of lamb. We admitted our mistakes.... And in the middle of a cold winter, there was birth and restoration and hope."
—naturalist Kevin Sweeney, from "Yellowstone Winter"

The campaign could fittingly be called "The Recall of the Wild." The restoration of America's most controversial predator, the wolf, had journeyed clear across the country, from its success in Yellowstone National Park in Wyoming to New York state, to support the proposed consideration of the release of perhaps fifteen to fifty gray wolves in the vast Adirondack State Park, just a few miles northwest of the state capital in Albany. But the Defenders of Wildlife and other environmental groups knew that the stakes would be higher in New York than in Wyoming; New York was known more for its urban jungles and dense human population rather than wild, open expanses and sprawling private lands. The last New York wolf was believed to have been killed in 1899 near Cranberry Lake in Adirondack State Park and, since that time, New Yorkers had, for the most part, seemed quite content to remain wolfless. How easy, then,

would it be to convince New Yorkers to restore the wolf into a public access land in the Adirondack region where, nearby, nearly a hundred and thirty thousand people lived? And how could small dairy farmers, whose farms ringed the park, be convinced that even though large packs of wolves could eat up to a ton of red meat each week, the fierce predators would not view the defenseless herds of placid animals as a sort of wolf McDonald's? New York's Farm Bureau certainly deflated the hopes of the Defenders of Wildlife that the New York campaign would be an easy one; the Bureau primed the audience before their arrival with warnings that the wolves would pose a threat to cows, horses, pets—even children. Perhaps the Farm Bureau hoped that such warnings would feed into the hysteria that had, long ago, initially branded all wolves as evil, making them pariahs on their own planet, hunted down more out of hatred rather than for sport.

Years ago, it had not been uncommon to inject strychnine into tallow or pieces of moose liver and then bury the poisoned treats where wolf tracks had last been seen. The wolves would then unknowingly dig up their last supper. Or the wolves would be herded by hunters out of the forest and into the open on a valley floor, where other hunters from airplanes would shoot them. The first method had been illegal, but largely ignored, and the second—aerial hunting—had not only been outlawed in nearly every American state, but more often than not sickened and enraged even the most stalwart wolf opponents. And yet many people today say they would resort to such methods if they could. Some hunters still see wolves as competitors for game in the wild. Some ranchers still see wolves as raiders of their chicken coops, barnyards, and pastures. But the philosophy of such individuals is not just to kill off all the wolves in ranching country and hunting country, but *everywhere*. They think that wolves should be shot with a gun and not a camera. They might grumble that there are plenty of other places for wolves to go—just keep them away from their own neighborhoods—or they may clearly and emphatically state that wolves simply have no business even being here, and vow to shoot any wolf they see.

Few people are indifferent to wolves; few remain neutral on the issue. Like the segregation issue that burned white-hot in the south in the 1960s, there are those who are so vehemently opposed to the reintroduction of wolves that they make no effort to hide their intent to use violence to oppose any wolf reintroduction. But rather than shout such people down or prepare to meet their threats of violence with retaliatory threats of violence, those who wish to defend the right of the wolves to live on the planet opt for more nonviolent methods, such as raising understanding through education and providing up-to-date information on American wolves in the wild. Wolf proponents have a dream, and their dream is that wolves and mankind can live in the same world, together, in peace and harmony. So they have become ambassadors who represent the wolf kingdom. Healthy wolves do not attack people, the wolf ambassadors tell their audiences. In Minnesota, home to two thousand gray wolves, only one wolf has come within twenty miles of St. Paul. The population of deer (currently numbering 71,000 in Adirondack State Park), elk, moose, bison, and other wildlife is now so abundant—which it was not earlier in the century, due to hunters' decimation—that the wolves will not have to resort to killing livestock or household pets for survival. (In fact, out of the one hundred thousand dollar fund set aside by the Defenders of Wildlife in 1987 to compensate ranchers in the Northwest for livestock damage from wolves, only $25,930 has been paid out for ninety-eight animals killed by wolves.) Wolves, the ambassadors further preach, are sociable creatures that travel in packs as families, not as marauding bands. Wolves are wary of human beings. Wolves are noble. And, as well, wolves are good for the economy; thousands of tourists have been drawn to Yellowstone's Lamar Valley, where a wolf pack often wanders through. "Let us prove all of this to you," the Defenders of Wildlife tell all parties who are concerned over wolf reintroduction programs in their area. "In fact, we will even pay for a study of the issue," they add, an offer which New York's Governor George Pataki has accepted.

Without question, the wolf has always occupied a special place in people's imaginations, from the Big Bad Wolf that destroyed the homes of at least two out of three little pigs (and then ate the two homeless pigs) and the cross-dressing demon that terrorized Little Red Riding Hood after devouring her grandmother to the majestic creature in Jack London's classic stories *Call of the Wild* and *White Fang* to the beast long revered by Native American cultures to the tail-wagging, yipping, four-legged creature that closely resembles humanity's beloved canine companion. Until recently, hatred and fear usually won out in any discussion of wolves; today, school children, who, many might think, would be scared out of their wits at the potential of an actual Big Bad Wolf sniffing around in the hopes of picking up their scent, following them home, and making a Happy Meal out of them, are some of the most receptive and vocal proponents of wolf restoration. When Kent Webber, founder of the Colorado-based Mission: Wolf, toured the Northeast with two wolves to build public support for the New York wolf restoration campaign,

he was astonished at the warm welcome the wolves received. School children mobbed his bus to get a glimpse of the wolves inside. Even though the traveling wolves, Sila and Merlin, are not trained, they are unusually social; they charge around the packed presentation rooms on leashes and rise up on hind legs to greet human friends.

The United States Fish and Wildlife Service identified the Adirondacks as suitable for wolf habitat in 1992, along with two sections of Maine and a small piece of northern New Hampshire, because such areas have enough deer, moose, and other prey for wolves to eat, few roads where wolves could be killed, and small human populations. With government support, along with packed Wolves in America conferences, nonprofit organizations such as Defenders of Wildlife, thousands of schoolteachers, countless wolf aficionados, environmentalists around the world, and receptive school children, it is only a question of time before the wolf population is restored in America to healthy— and more naturally and mutually beneficial—proportions.

Not every "nature exile" has experienced such an incredible turn-around in public sentiment and private support as the wolf; few of nature's "outcasts," which mankind has worked so hard over the years to be rid of, is ever so willingly welcomed back. The Austin, Texas, bats that live under the reconstructed city bridge are, like the wolves, a small exception to the rule; but, for the most part, the creatures that resemble flying rodents will probably never raise such nationwide public sentiment as the wolf, should its own species ever face extinction. Decades ago, when "vermin exterminators" boarded up all the exits in an abandoned bunkhouse in Coquitlam, British Columbia, and then released a killing cloud of cyanide gas, annihilating bats by the thousands—five thousand, to be

exact—newpaper headlines joyously proclaimed the demise of "this hot-bed of vampire life." Exterminators have, for years, earned an enviable profit eliminating bats and, even though the use of DDT was eventually banned, as recently as 1988 the government of Ontario issued special permits to exterminators to use the banned substance against bats. Even though bats do no structural damage when they are inside a home, the aroma of their accumulated droppings can be offensive and their very presence can be frightening. However, using poisons in bat control causes them to sicken and then fall to the ground, where they die slowly and, in the process, may be picked up by pets or children, which poses more of a threat to people and the environment than the bats ever did. (If bats are unwelcome tenants in your home, simply work on denying them continued access by blocking cracks and crevices in the spring or fall, when young or hibernating bats will not be inadvertently trapped inside, or at dusk, when the bats are out hunting. Remember, however, that unless you provide your ousted tenants with an alternative residence—a bat home or an old, hollow tree in which they can roost—you lose your natural and valuable controller of the crop-destroying insects in your yard, on which the missing bats once fed.)

Bats are certainly more tolerated today, sometimes even enticed onto a landowner's property with a bat house for the useful balance they help to maintain in a backyard ecosystem, but it is doubtful that the lowly bat, though a prized devourer of insects, will ever be viewed as noble or as majestic as the wolf or have its image displayed on key chains, bumper stickers, posters, coffee mugs, wallpaper, rugs, towels, and other merchandise, except during Halloween, when the image of the black bat—its wings outstretched, a vicious gleam in its beady black eyes, and its sharp white fangs poised to sink deep into soft human flesh—will always be a frightening reminder of the "witches' bird," an image it has never quite been able to shake.

The bat is a perfectly harmless and wonderfully beneficial creature. So how did it, along with so many other living things, ever become so unpopular, so disdained, so feared, so hated? For the most part, imagination and myth—not solid information—have helped to form the misconceptions. Despised species have been saddled with bad reputations that they have, unfortunately, usually never earned. Bats, mice, rats, spiders, snakes, toads, and wasps have all, at one time or another, been reputed to be killers, stranglers, disease-carriers, filth-lovers, evil-doers. They are deemed to be part of nature's "dark side"—incorrigible juvenile delinquents; unredeemable criminals; frightening monsters. Just think of all the Hollywood horror flicks you have ever seen and the creatures that "star" in such screamers. Werewolves. Tarantulas. Gila monsters. The Fly. Mountain lions. Sharks. Snakes. Rats. You will never see a panda bear holding a writhing Fay Wray in its paw as it clambers to the top of the Empire State Building while airplanes strafe the creature; instead, you will always equate gorillas with such awesome ferociousness. Mankind creates and then perpetuates such myths for its own benefit; because nature's outcasts repel rather than attract, because such living things are deemed by mankind to be objectionable, and because mankind has an intolerance for anything it finds objectionable, such creatures are placed on "Most Wanted" lists. They are hated, feared, despised.

Later, if or when mankind discovers that something it has sought to eliminate is actually harmless or even serves a useful, human-friendly purpose, mankind might begrudgingly give it space. But, for the most part, the motto is, *"Once an outcast, always an outcast."*

You may find it hard not to stomp on a spider, grind out the guts of an ant, club a snake, kick at a pigeon, snap the neck of a mouse, drown a beaver, swat at a fly, shoot at a raven, smoke out a woodchuck, or poison a slug. Because you define such things as pests, they cannot be tolerated. But, even further, because such things do not serve a useful purpose to you, there is, therefore, no need for them. The overriding feeling is that you must simply get rid of them rather than live with them, understand them, or look at them in a different light. Witness, for example, the slaughter of great buffalo herds; mankind did not need the buffalo the way Native Americans needed the buffalo— to the white man, the buffalo was not a source of food or clothing but a massive, slow-moving, dim-witted creature that took up valuable prairie space and impacted train schedules—so what difference did it make to shoot the peaceful beasts from the window of a passing train other than for the sheer sport of it? Witness, too, how the pigeon—today the bane of city parks, buildings, and revered monuments—was once considered to be vital to mankind and therefore treated with respect. Up to and during World War II, carrier pigeons were used by the military as unfailingly heroic messengers. Pigeons were also being trained as living guidance systems for anti-submarine missiles; the function of the three-pigeon crew, which was encased in the nose of a missile behind three little windows, was to guide the missile, kamikaze-style, to its target, where it would then burst in an explosion of fire and feathers. Even recently, the acute vision of pigeons, known to be appreciably better than human vision because the pigeon's beady little eyes see not only a stereoscopic image of objects that lie straight ahead of it but also can see in a wide-angle, monoscopic periphery, were used in a United States Coast Guard research program known as Project Sea Hunt. "The idea," writes Des Kennedy in his book *Nature's Outcasts: A New Look at Living Things We Love to Hate,* "was to exploit the pigeon's acute sight and its learning ability in order to spot shipwreck survivors at sea. A three-pigeon squad of ordinary street pigeons was trained to peck a key upon spotting an orange life jacket. In eighty-nine trial flights the pigeons, suspended in a transparent box beneath the helicopter, spotted an orange buoy 96 percent of the time, sometimes from as far away as seven hundred meters. The human flight crew, by comparison, had a meagre 35 percent success rate. Though promising, the project was eventually scrapped due to budget cutbacks."

Thus retired from any "active duty" that made use of their skills or spotlighted their interdependent benefits to mankind, pigeons today flock in great numbers in cities, where their only known predator, besides hawks that have been reintroduced into the cities and suburban owls, is man. A small Pennsylvania town holds an annual pigeon shoot where the caged birds are released and then shot by paying customers. Chicago established a program of trapping and gassing ten thousand pigeons a year. New York City dabbled with a contraceptive program before it turned the matter over to a dozen pairs of peregrine falcons. San Francisco placed fake owls around the city to terrify its pigeons; after a short, respectful period in which they kept their distance, the pigeons later took to perching on the owls and streaking them with droppings.

The gentle birds that strut and coo and scramble for scraps in municipalities are not just "rats with wings," as actor, director, and New Yorker Woody Allen once called them; they are intelligent scavengers. They, like every other living thing, play a vital role in balancing the ecosystem. In European cultures, many are bred and pampered as show pigeons, homing, or racing pigeons. In Hong Kong, the bird is a deep-fried delicacy. In America, pigeons have played a vital role in military history and the defense of American freedom. Simply slaughtering them by the billions is not the way to handle their great numbers, no matter how you may feel about them. Mankind has already done irreparable damage in singlehandedly causing the loss of one of the pigeon kingdom's more valuable species—the swift, magnificent, unerringly determined passenger pigeon—which was hunted down in the name of sport and for pigeon pie. The last of the passenger pigeons died in a Cincinnati zoo in 1914, owing its extinction to mankind's intolerance and ignorance of the outcasts.

Mankind does not just consider nature's outcasts to be those in the critter category—hyenas, vultures, pigeons; those in the insect world—mosquitoes, fleas, flies; those in the mollusk world—snails, slugs, horseshoe crabs; those in the reptilian world—snakes, lizards, alligators; or those in the arachnid world—spiders, ticks, scorpions. Other natural living things despised by mankind are not creepy, but creeping: moss, alder trees, nettles, dandelions, ragweed, pigweed, touch-me-not, violets, clover, smartweed, goldenrod, poison ivy, pokeweed, milkweed, crabgrass, bindweed, and countless other fast-growing

Maine, 1958 © David Vestal 1997

living things on which mankind spends a fortune to kill or, at best, keep at bay from its well-manicured lawns, fruit orchards, and vegetable and flower gardens. And yet, in creating such Eden-like plots, mankind attracts the very things that it despises—the weed seeds—which would much prefer to live in a plot of enriched soil or a field or meadow.

When, in 1870, Frank J. Scott sought to bring respectability to the middle-class city people and published the first volume ever devoted to how to live in suburbia, *The Art of Beautifying Suburban Home Grounds,* the suburban landscape of America was forever determined from that point on. "A smooth, closely-shaven surface of grass," wrote Scott, "is by far the most essential element of beauty on the grounds of a suburban house." One's lawn, determined Scott, said much about the occupant, so it was vital to maintain a lawn that had excellent quality. "Let your lawn be your home's velvet robe," urged

Scott, "and your flowers its not too promiscuous decoration." To not maintain one's lawn to such high standards was, Scott warned, "selfish, unneighborly, unchristian, and undemocratic." Thus, green living things became instantly separated into two categories—those that were acceptable and those that were unacceptable. Mankind began to feverishly stock sheds, garages, and basements with all sorts of mechanical equipment and chemical formulas to maintain such suburban single-mindedness: weedwhackers, lawnmowers, tillers, hoes, shovels, hoses, sprays, dusts, granules. And thus was born a seasonal ritual which, on a weekly basis, pitted man against nature. Because the suburban dream was to be able to look out of one's living room window and see a sea of green all the way to the horizon, trees were cut down. Anything that would obstruct the view of The Green Ocean was removed; in fact, the erection of a fence cast suspicion and doubt over the patriotism of one's neighbor. Mowing became a tract development ritual—a gathering together of men and their mowers, a powerful proclamation that, in presenting a noisy, sharp-bladed, united front, mankind could, with its protective moat of lawn, defend itself against any natural infiltration of its beloved turf.

Today, lawn care is viewed as an important civic responsibility in the suburbs. This means not just maintaining well-trimmed greenery—after all, clover and buttercups and violets and dandelions and crabgrass all *look green* when they are trimmed to a certain height—but maintaining a well-trimmed, *single variety* of greenery. Suburban dwellers have thus learned how to dust the dandelions, the best ways in which to uproot the turf invaders, and, when all else fails, how to resod and reseed until the lawn is completely purged of weeds and meticulously coiffed. "Nowhere in the world are lawns as prized as in America," writes Michael Pollan in *Second Nature*. "In little more than a century, we've rolled a green mantle of it across the continent, with scant thought to the local conditions or expense. America has some fifty thousand square *miles* of lawn under cultivation, on which we spend an estimated \$30 billion a year."

In many communities, lawn care is even determined by law. Not a summer season goes by without some controversy erupting in a suburban community over the failure of a single homeowner to mow his or her lawn. "Not long ago," recounts Pollan, "a couple that had moved to a \$440,000 home in Potomac, Maryland, got behind in their lawn care and promptly found themselves pariahs in their new community. A note from a neighbor, anonymous and scrawled vigilante-style, appeared in their mailbox: 'Please, cut your lawn. It is a disgrace to the entire neighborhood.' ... Some [neighbors] offered to lend the couple a lawn mower. Others complained to county authorities, until the offenders were hauled into court for violating a local ordinance under which any weed more than twelve inches tall is presumed to be 'a menace to public health.'"

Even though laws of this kind are on the books in hundreds of American municipalities, just who is it that determines what constitutes a weed? Who determines what living thing has a right to put down roots and what living thing should be uprooted? Just because there are some that might prefer to look out of their living room window and see a sea of grass does not mean that those who wish to look out of their living room window and see a field of goldenrod or clover cannot do so; each party considers their view to be the more beautiful, peaceful, harmonious, and virtuous. "What is a weed?" American essayist and poet Ralph Waldo Emerson once asked. His answer: "A plant whose virtues have not yet been discovered." Perhaps Emerson is not the

only person to feel such acceptance of what others might deem deplorable. Case in point: a man in a suburb of Buffalo, New York, has spent the last several years in court—and nearly twenty-five thousand dollars—defending his right to grow a wildflower meadow in his front yard, much to the dismay and disapproval of his neighbors, who obviously do not share in his enthusiasm. When the neighbors mowed down the offending meadow, the man erected a sign that said: "This yard is not an example of sloth. It is a natural yard, growing the way God intended." A local judge disagreed with the man's interpretation of natural growth and ordered him to mow his lawn—or face a fine of fifty dollars a day. Thus far, the man has opted to pay the price for enjoying his meadow, welcoming all the living things that have been cast out by his neighbors.

If the pristine lawn is a sacrament, then the Buffalo man—

and many others—are sinners. But just as there is no "right" religion, no "right" way of worshipping, no "right" philosophy of belief, and no "right" spiritual essence, so, too, are there no "right" or "wrong" living things. There is a reason for the existence of every living thing; mankind has only to eliminate one species to learn this simple lesson. For example, the deer population, once kept in check by wolves and coyotes, has now grown to such huge proportions without its natural predators that, in many areas, the proliferation of deer has created an ecological disaster. Because deer browse on all kinds of foliage, they decimate the new growth of Douglas fir seedlings and, their favorite winter food, western red cedar seedlings, much to the dismay of foresters. Safe from wild predators and hunters, deer populations have also increased dramatically in suburban areas, eating whatever humans have planted for their own use. Because of this, some today now view deer as pests—not as the gentle, beautiful, Bambi-esque creatures they once thought. Which living things you choose to consider as outcasts depends on how a particular living thing has impacted you, your environment, and your life. And this impact is, more often than not, a direct result of the actions mankind takes because of its own particular view of what constitutes an outcast.

Show an adult a snake, and more often than not the adult will recoil in horror, fear, or loathing. Show a child a snake, however, and the child will look at it with curiosity and wonder. The child has not yet learned what the adult may have learned from experience (by being bitten by a snake, for example), from an adult or authority figure (such as a parent who communicates a fear of snakes), or from a story or movie that featured the snake in a frightening or horrifying way (such as a rattlesnake that startles a Western hero's horse). When a Stoneham, Massachusetts, neighborhood learned that a neighbor's deadly banded Egyptian pet cobra disappeared when its owner had taken it out to the front yard to sunbathe, police distributed "Beware Missing Pet Snake" fliers door to door, knots of nervous parents kept watch over their children, and school officials at the nearby elementary school met to institute a plan to limit school yard access. A few months later, one of elementary school students came across the snake in his classroom and alerted his teacher. When asked later by a reporter if he had been frightened when he found the snake, the little boy replied, "Nah. I thought the snake was really neat-looking and it was just scared of everybody."

Horror, fear, and loathing of any living thing are things that are taught or experienced firsthand and then perpetuated not just by continuing to believe in the original lesson but also by refusing to revisit or rethink what has been learned. Being stung by a bee or mauled by a bear is certainly painful and life-threatening. But does that mean that all bees and all bears will do the same thing? Does that mean that all bees and bears must be eliminated? "Why did the bee sting?" or "Why did the bear attack?" are the more important questions to ask, for it is then that you learn about the interrelationship between all living things and why creatures do the things they do.

When an old man who lived on a farm just outside of Bonnechere, Ontario, stepped out onto his front porch one morning and saw a wolf in his field, he immediately got his gun and then shot the wolf. "It wasn't bothering anything," the man admitted, adding that he had, long ago, stopped raising poultry or livestock. "But I want to get even. I know what they can do." Since his farm had been raided by wolves in previous decades, when the farm was a working farm, to him *all* wolves were unacceptable. "I'd shoot them every time," he said in a determined voice, choosing not to listen to how his family's farm, which had been built in 1917, had encroached on wolf territory or how the dwindling population of deer, beaver, and moose that had been hunted down had made the wolves turn to the man's sheep and calves. "Wolves kill livestock and take away my family's money," was all the man had learned. And, because of this, his opinion would never change. "I know wolves," the man narrowed his eyes and squinted into the forest that surrounded his home. "They got no business being here."

"Tread lightly upon this little piece of earth," it has been said as a reminder of the potentially devastating impact mankind can have upon the planet and its ecosystem. Because mankind is not the center of the universe—merely a part of the web of life—mankind's success as a species rests upon its ability not to defeat those living things that it identifies as enemies but to live effectively with *all* organisms—from the tiny flea to the towering elephant. Ironically, some of the species most despised by mankind are those that most closely resemble mankind; they, like mankind, are competitive, persistent, tough, aggressive, adaptable. They, like mankind, can overpower other species, grab more for themselves, multiply even under the most degrading and desolate conditions, annihilate. A sting from a bee can kill, a speck of oil from a poison ivy leaf can make even the strongest man slave to salves and ointments, ivy can crumble bricks and mortar, and fleas can kill in multitudes—twenty-five million people in Europe alone died from the Black Death that the lowly fleas single-handedly spread across the continent. And yet each living thing keeps other living things in check so that no one living thing,

including man, becomes too powerful or too proliferate.

The greatest challenge that mankind faces today is reconnecting with the natural world, including those parts of it that mankind has learned to fear and despise. Mankind must cast off careless assumptions and time-twisted superstitions, for beyond them lie the truth. There needs to be a wholesale change in attitude towards all living things—a clean-slate approach in which what was once believed is now challenged for the purpose of recreating a world of healthy diversity and gentle tolerance. The words *insignificant* and *inconsequential* ought never again to enter into any discussion of a living thing. All living things are of much more consequence, have much more influence in the economy of nature, and are much more mighty in their effect, whether they themselves are minute or massive, than mankind realizes.

Some Ways to Embrace Nature's Outcasts

1. Read books and articles or watch videos about a particular natural living thing that you fear or despise. Find out how it lives and what purpose it fulfills as a part of the ecosystem. The next time you come in contact with this living thing, try to respond to it differently—even if it still sends chills up your spine!

2. Keep in mind that most living things are more afraid of you than you are of them, even those creatures that have the power to make you ill with their poisons, severely injure you in an attack, or even kill you. However, if you are going to be hiking, camping, swimming, or participating in any other outdoor activity that might bring you in contact with a potentially dangerous creature, abide by posted warnings. Be sure you know the rules of behavior that most living things respect. Practice "survival drills" before you set out. Pack any

lifesaving gear or medical supplies you might need. Never feed any wild animal. And, if it is not wise for you to venture out at a particular time, such as when bear sows and their cubs are particularly active, postpone your planned activity.

3. Keep in mind that you are always the interloper, no matter where you live. That means that creatures you may fear or despise live there, too. Just as they must tolerate your presence, so, too, must you tolerate theirs.

4. Because you may be entranced with nature's more majestic wonders—lions, giraffes, orchids, redwoods—it may be easier for you to consider less majestic living things as outcasts: ants, squirrels, earthworms, daisies. Yet such common and less obvious living things are every bit as remarkable, elegant, graceful, and, in their own unique way, beautiful. Expand your range of love for the many species that inhabit the earth.

5. French writer and philosopher Voltaire once wrote, "Ask a toad what is beauty—the great beauty, the *to kalon;* he will answer

that it is the female with two great round eyes coming out of her little head, her large flat mouth, her yellow belly, and brown back...." Strive to see one of your "outcasts" in the same way a member of its own species—preferably of the opposite sex—might see it.

6. Learn the names of the weeds in your garden. Make a bouquet out of flowering wild weeds. Closely examine the clover in your lawn; try to find a four-leaf clover. Observe how bees and butterflies are as attracted to the flowering weeds on your property as they are to the flowers.

7. Ask your children or your nieces and nephews if they are frightened of or dislike any living thing. Encourage them to talk about why they feel the way they do. Then, use a trip to a wildlife sanctuary, science museum, or petting zoo or a book you read together to help ease their fears or change their attitude toward one of their "outcasts."

8. Become an "outcast observer" by allowing an "outcast" that cannot

harm your environment or endanger people or your household pets to reside in or on your property. Observe its comings and goings. Is it nocturnal—active at night—or diurnal—active during the day? What does it eat? What is its "home" like? Does it follow a pattern from day to day? Does it change itself or its behavior in any way?

9. Find out how your particular "outcast" interacts with other species. Discover what living things might consider your outcast to be an outcast, too, and why. What other living things have an interdependent relationship with your outcast? What would happen to other species if your outcast became extinct?

10. Remember that even in the most man-made environment, inhabitation of other living things can occur. Rather than resort to chemical means of extermination in your apartment, home, garden, garage, basement, or attic, seek first to provide natural controls. Seal potential points of entry.

Repair rotted wood and broken window panes. Keep foods in airtight containers or tightly wrapped; if necessary, freeze flour, cereals, and other grains larvae-producing insects love. Regularly clean underneath bird cages and in the areas in which you feed your pets. Change the water in bird baths on a daily basis. Keep gutters free of standing water and rotting leaves. Position bird feeders away from your house in order to prevent squirrels from seeking shelter close to a food source. Ask your chimney cleaning service to install a grate over the top of the chimney to keep out nesting creatures. Use tight-fitting lids on garbage containers and, when necessary, place rocks or cinderblocks on top to prevent wild animals from foraging through your trash. At night, bring in any dog bones or dry cat or dog food. While such measures may not eliminate unwanted outcasts from invading your home or property, they are effective, safe, and inexpensive controls.

Beastly Behaviors:
Promoting More Natural Captivity

"A tiger who cannot hunt, cannot mate with other tigers, and cannot explore and survey its territory has little chance of feeling pride. Perhaps, for some tigers, a chance to jump through a flaming hoop is better than nothing. But should tigers have to settle for something that is better than nothing? To turn these magnificent animals into slaves, and then degrade them further by making them perform tricks for human amusement shows as much about human abasement as it does about animal capabilities. That a tiger is condemned to slow death by boredom unless it finds pleasure in performing is a sad commentary on what humans have done to these magnificent predators."
—psychologist and writer Jeffrey Moussaieff Masson and biologist and journalist Susan McCarthy, from *When Elephants Weep: The Emotional Lives of Animals*

"... animals feel anger, fear, love, joy, shame, compassion, and loneliness to a degree that you will not find outside the pages of fiction or fable. Perhaps this will affect not only the way you think about animals, but how you treat them."
—Jeffrey Moussaieff Masson

It was a scene that stunned those in charge of the transfer of their living cargo—a seven thousand pound ailing Orca that had lived for decades in captivity in a tiny, manmade enclosure, taught to perform tricks for mankind and made so dependent upon man that it could not eat unless it was hand-fed by trainers—a stunning scene that they thought could only be reserved for the fantasy world of movies, never for real life. It was after two o'clock in the morning, a time when most people were at home, fast asleep in their beds, but at this early morning hour on this particular day, thousands of people in a Mexican city were wide awake. They lined the streets, all the way from the Reino Aventura amusement park to the airport, where a cargo transport plane awaited its killer whale passenger. Men openly wept. Families hugged one another. Little boys and little girls cried and waved as an enormous container with its equally enormous cargo lumbered by. Then, as if on some sort of innate cue, the people joined together and formed a spontaneous parade that followed the entourage of police vehicles with their wailing sirens and flashing lights, support vans, and utility vehicles that were carrying scientists, equipment,

and dozens of workers who were needed for the massive relocation project. The Mexican people danced and sang and laughed and cried; as difficult as it was for them to say goodbye to their beloved whale, they knew that they must celebrate in this journey of the whale—a journey that they and so many others had grown to realize was *the right thing to do*. For even though the journey would take their magnificent and profitable whale away from them, it would bring the whale closer to his own sense of freedom. And that, they knew, was of far greater importance.

Reino Aventura had not captured the whale; it had originally purchased him from another amusement park, thereby perpetuating the whale's life of captivity. At Reino Aventura, the whale quickly become the star attraction and drew enormous crowds. But the size of the crowds and the scope of the whale's fame could not compare to what happened after he landed the title role in Warner

Brothers' hit movie, *Free Willy*. For, after 1992, Keiko—who had, overnight, become the most famous Orca in the world—had also become the most visible and moving symbol of evolutionary thinking regarding animals in captivity. International attention focused not only on the tragedies in his life story, but on the role humanity had single-handedly played in preventing the young whale from living a peaceful life of social harmony and cooperation with other members of his family—a life in which he would have probably lived nearly half a century in wild and exhilarating oceanic freedom—and succeeded in making him into a spectacle that, day after day, suffered physically and emotionally. The plight of Keiko the killer whale and the ensuing formation of the Free Willy Foundation revealed that mankind was now ready to react dramatically and with determination to convert its own view of itself, its own human nature, from a position of cruelty and greed in its treatment of animals to one in which responsible actions could be taken in the welfare of wildlife that were based on compassion and love. Mankind was ready to make amends for all the animals that had ever been held in captivity for human amusement and, of course, for profit; Keiko became the poster child of this movement. And, until Keiko was finally able to embark on the first leg of his eventual journey home, thousands protested the long and politically charged process of "prisoner" release that had to be negotiated between the Mexican and American governments. The people pushed and pushed, wrote letters, spoke out, called hotlines, held fundraisers, and prominently displayed signs that read, *Keiko is Still a Prisoner*. Long years later, when the plane carrying Keiko finally touched down on the first leg of his journey toward freedom in the small seacoast town of Newport, Oregon, joyous residents who lined the streets cheered, clapped, cried, waved American flags, and held up signs that read *We ♥ Keiko*.

The process of bringing Keiko home focused a harsh spotlight on humanity's practice of keeping animals captive for decades and painfully revealed that what had begun as a simple, thoughtless capture for profit, like so many others that had been executed over the years, was one that turned out to be not so simple to correct and was unbelievably cruel and inhumane, even under the best of conditions and with the care and guidance of the most gentle, kind, and loving trainers.

In the frigid North Atlantic waters off the coast of Iceland, at least seven different pods—or families—of Orcas have been identified through the complex and distinctive language of whistles, clicks, and cries that they use to communicate with one another, as familially distinctive as your own family crest, genetic traits, or rituals and traditions. In fact, each pod's dialect is so unique that a member of one pod can recognize another member from miles away. Keiko learned the language of his pod right from his birth around 1976 and practiced it often with other members of his family, for whales live all of their lives with their family of origin. At the heart of an Orca pod is the mother; brothers, sisters, and even the maternal grandmother stay together in this permanent, closely bonded relationship. Keiko most likely nursed for almost eighteen months before his mother began to teach him how to hunt schools of herring, the favorite food of the Orca. But he was probably too young and inexperienced to learn how to avoid the nets of fishermen who competed with the whales for herring. When Keiko became entangled in the fishing nets, the fishing captain was obligated to cut him loose. But because the captain had been offered thousands of dollars to capture a young Orca, a creature which is highly trainable at an early age and therefore valuable to marine amusement parks, he made a choice that was influenced not by a moral obligation, but by his own financial benefit. Keiko was not set free.

Keiko survived the frightening ordeal of his capture and was shipped from Iceland to Marineland in Ontario, where he joined six other Orcas that were being trained for sale to amusement parks. But because Keiko was the youngest and the smallest, he was constantly harassed by the older, bigger whales. He became timid and less willing to perform; he remained in Canada for five more years because no other amusement park wanted a nonperforming whale.

In 1985, Marineland sold Keiko to Reino Aventura amusement park in Mexico for one hundred thousand dollars. Because Keiko was the only Orca at the park and possibly because of the unique bond he quickly formed with his two patient and loving trainers, Keiko soon reveled in the spotlight. He became the star attraction and drew enormous crowds over a seven-year period, until 1992, when Warner Brothers began to look for a whale to star in the movie *Free Willy.* The film became a hit not only at the box office, but also in the hearts of audiences who were moved by the plight of such a graceful and intelligent animal being relegated to a lifetime in captivity. As the film credits rolled on the screen at the end of the movie, viewers were informed that they could call 1-800-4-WHALES to learn more about whales. Nearly three hundred thousand calls were logged *a month;* what callers were most concerned about was not so much the plight of *all* whales but the plight of the real-life Willy— Keiko—who, at the end of the movie, did not earn his freedom but, instead, returned to his adoptive home in Mexico to resume his life of servitude. Movie-goers by the thousands conveyed their outrage and dissatisfaction that life, in this case, was not imitating art.

As a result of public sentiment, Warner Brothers began to work with Dave Phillips, an experienced whale protection activist, to devise a plan that would help to free Keiko. Children around the world became the driving force in the decision to do the right thing for Keiko as they put enormous pressure on Reino Aventura. After two long years of negotiations, the amusement park finally agreed to donate Keiko to the Free Willy Foundation. Once Keiko's ownership was transferred to the Foundation, the search was on for a new site for Keiko. But a complete physical examination of Keiko revealed that despite excellent care from his trainers, his condition was rapidly deteriorating. Because the size of his pool was too small, his dorsal fin had atrophied due to lack of exercise. He also had a weakened immune system. Of greater concern, however, was his pronounced skin rash, caused by a virus aggravated by chlorinated, artificial salt water

kept at temperatures comfortable for his trainers to join him in the pool but far too warm for him. Because it was feared his virus might be contagious to other marine mammals, that not only meant Keiko lost the option to be transferred to another facility, but that a facility had to be *created* for him. That, in turn, meant that a large amount of money had to be raised and Keiko had to stay longer at Reino Aventura in an environment that was slowly compromising his health.

Foundation members frantically searched a variety of coastal sites on which to build the first whale and dolphin rescue center, where Keiko would be the first whale tenant and where the emphasis would be on rehabilitation, not on performance or entertainment. After much laborious research, the Oregon Coast Aquarium near the town of Newport was selected to be the site of Keiko's new home. Seven million dollars was donated, a hundred thousand dollars of which was raised by school children through drives as simple as collecting change in schoolroom piggybanks. Because Keiko needed to be transferred as quickly as possible, construction that would have normally taken up to two years to complete was fast-tracked to eleven months by using a design-build process; as soon as a design was completed for a particular section of Keiko's facility, the building immediately began. Keiko's trainers in Mexico began to integrate a new "act" with Keiko by training him to grow accustomed to being placed in a huge canvas sling and lifted out of the water, an activity he would need to experience once when leaving Mexico and again when arriving in Newport. A marine mammal veterinarian with experience in whale transportation was assigned the task of choreographing Keiko's transfer. A C-130 cargo plane was donated by the United Parcel Service (UPS ♥ Keiko!). A massive construction crane was obtained to lift the fifteen-ton container in which Keiko would ride for nearly fourteen hours. Three thousand pounds of ice essential to maintaining Keiko's water container temperature were readied.

On the last day of his performing life, Keiko leaped and splashed in front of sold-out audiences. At the end of his final routine, he slowly and gracefully leaned his massive body to one size in his small pool and waved a flipper in the air, his gesture of farewell to a grateful audience; in return, a sea of Mexican children and adults waved back to him with their own heartfelt farewell. Keiko's stage life had come to an end.

At midnight, Keiko was fitted into his customized canvas sling, lifted out of the warm tank water, and hand-dried by his trainers, who then rubbed him with moisturizing cream to keep his skin from drying out. Under the glare of bright lights and the roar of the crane's deafening diesel engines, he was then lifted several feet above the ground and slowly lowered into a container filled with cold water and ice.

The unanticipated numbers of spectators who had lined the streets to say goodbye to their most famous resident, although well-meaning, actually jeopardized the journey by adding two hours to Keiko's already lengthy and stressful relocation schedule. Then, when Keiko finally arrived at the airport and his container was placed on rollers to slide into the belly of the plane, one of the rollers suddenly jammed. It took another two hours to free the jammed roller. The additional time melted all the ice and compromised the temperature of the water in the container; more time had to be allowed during a scheduled refueling stop in Phoenix for the water to be changed and two thousand pounds of ice to be added. Then, en route to Newport, the flight crew was informed that bad weather threatened to close the airport. Keiko's journey to freedom became a frantic race against the wind and rain on the Oregon coast.

But then a break in the weather, a hasty clearance for landing—and the plane touched down safely in Newport. Once again, hordes of people greeted Keiko. Reporters from around the world and thousands of spectators who had waited hours in the rain lined the streets to greet Keiko. The entire world watched the event via satellite as Keiko was slowly lifted out of his container, suspended forty-five feet above ground, then lowered into the natural sea water in his new home. With a splash from his fluke, Keiko signaled his delight in his new home—a space five times bigger than his tank in Mexico. The greater size allowed Keiko for the first time in over fifteen years to dive and leap. Constant swimming soon built up his strength, stamina, and energy. His appetite increased; he began to eat twice as much and, in six months, gained one thousand pounds. His flippers increased six inches in width. He grew one foot in length. His skin virus completely healed. Of greater delight, however, was that in his new environment—one that more closely replicated his natural one and freed him from the stress of performing—Keiko, in a rare display, suddenly began to vocalize in his Orcan dialect.

Although Keiko must still learn how to survive in the wild by being retrained to "think" like a killer whale so he can hunt, and, once again, politics must be overcome before undertaking the last leg of his journey home to Iceland, Keiko's newfound desire to speak is what will ultimately provide him with the greatest assistance in his eventual release into the wild. For, through analysis of his vocalization structure, he can then be set free near a pod that "speaks" a similar dialect in the hopes that he will, in the near future, be reunited with his family of origin in the North Atlantic.

"Why can't Keiko find a new home?" was the question most often asked by school children during their participation in the campaign to free Keiko. But the question that was rarely, if ever, asked was, "Why did Keiko have to leave his home to begin with?" Captivity is, by definition, about restraint, control, loss of freedom, confinement, imprisonment, slavery; captivity, too, is solely a human activity, an action that is exerted only by man over beast (never beast over man) or by man over fellow man. In human society captivity serves a useful purpose when it imprisons rapists, murderers, robbers, drug dealers, child molesters, arsonists, and abusers; captivity removes such individuals from the normal ebb and flow of society and prevents them from continuing to harm others, thus protecting lives, property, and possessions. But it is not always the guilty or the harmful who are forced into such captivity; so, too, are innocent, harmless people placed in captivity when they disagree with politics, inadvertently become caught in the struggle between two cultures or conflicts between nations, hold and value beliefs different from those of the majority, or are viewed to be inferior because of the color of their skin.

Yet wild animals fall into none of these categories. Wildlife was first placed in captivity thousands of years ago by royalty who used their own human captives to create exotic menageries as a way of showing power and wealth. Queen Hatshepsut of Egypt, for instance, once arranged to have a giraffe shipped fifteen hundred miles down the Nile River as a way of conveying her ability to get whatever she wanted. Centuries later, when the public wanted to form their own menageries for display, they created zoos.

When zoos first opened their doors to the public, the goal was to have one example of as many animals as possible not only to build community pride, but also to foster a competitive spirit with other communities. Wildlife collections and rare species were guarded jealously; collecting trips to Africa and Asia were done in secrecy. Little care was taken in how a particular wild creature was captured or how it was transported to a particular zoo. If a creature died en route of disease or stress or as a result of being confined for days in a cramped cage without food or water, so be it; another animal could easily be obtained, for the wild was teeming with replacements. Too, little attention was paid to conditions of prolonged captivity at zoos; visitors would stroll and gawk at the imprisoned beasts and then leave, not realizing that most creatures received little more than low-maintenance care.

Then, in the early 1900s, Carl Hagenbeck of Germany created a sort of "zoo revolution" when he did away with the cages and allowed his animals to roam on wide green lawns surrounded by hidden moats. People flocked to his zoo, eager to experience this more natural view of wildlife, but it took several years for other zoos around the world to follow suit. Because zookeepers found it hard to control diseases that spread in the soil and grass of outdoor exhibits and cement boxes were far easier to keep clean through a simple and quick daily hosing, animal and zoo advocates mistakingly claimed that it was better in the long run for the animal's health to live in such a cement environment. It was only after World War II that some zoos began to feature Hagenbeck-like exhibits for certain wildlife, such as African cats and hoofed animals.

But, for the most part, the zoos that maintained the barred-cage exhibits slowly began to lose their appeal. Visitors no longer wanted to look at panting, perpetually pacing jaguars or bored, listless, overheated polar bears. Such animal captivity began to provoke pity and guilt among zoo visitors. Rather than be subjected to such deplorable conditions, the visitors simply opted to stay away. Zoos began to lose funding, more and more animals began to die in captivity, and zoos became the bane rather than the boon to communities.

Then, in the 1960s, animal lovers and activists began to champion the rights of animals, thus giving birth to the development of an animal-rights consciousness movement. Over the years, as the consciousness matured and more and more people began to listen, to speak out, and to publish research, the public began to question not just the credibility of zoos, but the role they should actually fulfill. Many people even went so far as to ask if there was a need for zoos.

But as the debate raged on about the ethics of capture in the wild and holding captive animals in zoos, and as many zoos and wild animal "farms" and "parks" began to close their doors and either relocate their captive wildlife to other zoos or euthanize them, scientists then stepped in and provided shocking evidence that immediately silenced advocates on both sides of the issue. Some species in zoos, the scientists revealed, were barely surviving in the wild. In some cases, in fact, certain species that were being held in zoos *outnumbered* their species in the wild. Captive animals suddenly changed from expensive headaches to precious possessions. As wildlife populations outside the zoo began to shrink, zoo directors and conservationists around the world promptly drafted a new role for zoos. They first determined that the days of collecting wild specimens were over; instead, zoos needed to replenish themselves and thus began to experiment with breeding. By the mid-1980s, more than ninety percent of zoo animals were bred in captivity while, in their own wild habitats, numbers of species were still declining on a daily basis as they continued to lose the battle not just against poachers but against encroachers—humans who were overtaking their habitats or degrading the quality of their habitats through deforestation, pollution, and development.

"Extinctions," writes wildlife expert Janine M. Benyus, author of *Beastly Behaviors: A Watcher's Guide to How Animals Act and Why*, "are occurring at a rate unprecedented in the planet's history, rising from a loss of one species every five years in 1850 to a current estimated rate of one species an hour, or nearly 44,000 species lost every five years. You may notice the proliferation of Vanishing Species signs in front of the animal exhibits at your zoo. The prognosis is not good for many of these faltering species; experts predict that, within the decade, the last of the last will perish from the wild forever." What this means is that zoos have now become modern-day "Noah's Arks"—saviors of species and, in some cases, the sole protectors of entire evolutionary lines. The role of zoos today is not just to protect and breed animals to remain in captivity, but to protect and breed animals for eventual release back into the wild into suitable habitats. This means that zoos have not just become valuable safety nets, centers for research, and places in which the public can be educated about animals and their behaviors, but, as well, safe habitats in which "wild" behaviors can be preserved, protected, and promoted.

Today, the best zoos are the ones that allow their inhabitants, through wildlife-like habitats created for the animals and superior care, to replicate their behaviors in the wild so wildlife can be born, grow, learn, resolve conflicts, socialize with members of its own species, build a home, win a mate, parent, and age. Such behaviors are essential not only for those animals that will eventually be released into the wild so they can survive even after a captive-bred existence, but also so behaviorists can get a glimpse of animal behaviors that would normally be difficult or nearly impossible to study in the wild. For example, it was in a zoo setting that the facial expressions of wolves were first studied in detail; too, the solitary nature of the panda and its remote habitats had long prevented researchers from learning about its reproduction until zoo pandas were observed mating and the first panda cubs were born in zoos. The chance to study live animals at a close range makes zoo study not just fascinating and revealing, but, in the long run, lifesaving to animals in the wild. By knowing more about how animals behave, conservationists and environmentalists have more information to use in going up against logging companies to prevent clear-cutting, in safeguarding wetlands, in preserving open, undeveloped areas, and in protecting the habitats of already endangered species or those that could be potentially endangered. While there are limitations to what can be studied within the confines of a zoo—for example, the impact of certain diseases on a population, migratory habitats, and predatory-prey relationships are some things that cannot be observed—zoo animals can teach zookeepers much about themselves and their everyday routines, thereby making zoo-based studies valuable in providing baseline data for field studies.

Yet despite overwhelming evidence and increased public sentiment, much, unfortunately, can still be written about the sad and shocking decline of certain once-revered zoos as well as the inhumane and horrible treatment that continues to go on with creatures being held in tragic examples of antiquated captivity. Take, for example, the case of the Stone Zoo, located just outside Boston, where Major, the polar bear, has spent twenty-two out of thirty years in a small, spare, dull concrete pavilion with a pool hardly big enough for him to swim in. A *Boston Globe* writer described his view of "poor Major" as "...the very picture of ursine apathy, he's like the last prisoner at Alcatraz, doing hard time at the

Stone.... It's a disgrace to keep an animal this way, and zoo officials admit as much." In his article titled "Let Them Go!" reporter Scot Lehigh also describes the monkey exhibit at the zoo. The creatures must share their outdoor quarters with the white-nosed coati, a Central and South American raccoon relative. Since the two species are incompatible, on the days one is on exhibit, the other must stay indoors. And, a few exhibits away, a black bear being held on a temporary basis for the state Division of Fisheries and Wildlife is on display "...in another dull animal prison. This sign hangs in the viewing area: 'By today's standards, this area is far from optimal for exhibiting wild animals.'" And yet, minute after minute, hour after hour, day after day, the animals endure their torture with the full knowledge of zoo officials, the city government, and the citizens of the state of Massachusetts.

When, in 1984, both the Atlanta Zoo and the Stone Zoo

landed on a list of the nation's worst, the residents of Atlanta had reason to be greatly ashamed of the conditions at their own zoo. The Atlanta Zoo housed its apes, lions, and tigers in "cramped dungeons"; incompetent keepers secretly turned over an elephant to a traveling circus, where she died; prairie dogs were mistakingly cemented in their burrows; a custodian was accused of eating rabbits from the petting zoo; the polar bear had only crushed ice on his concrete floor to keep it cool in the hot, humid southern climate; and Atlanta's own "Major problem"—a gorilla named Willie B.—was housed alone in an oversized bathroom with only a tire swing for diversion, so apathetic that he would stare blankly into space, head on his hand, as people filed by. Atlanta Zoo director Steve Dobbs said there was little he could do. "I'm far too removed from the animals," he told a reporter from *Parade* magazine. "They're the last thing I worry about, with all the other problems."

Unlike Boston, however, Atlanta took its zoo problems to heart and shamed itself into taking effective action. Twelve years later, the Boston zoo continues to deteriorate. But the new Zoo Atlanta is widely regarded as one of the ten best zoos in the country. In one of the most dramatic turnarounds in American zoo history, the antiquated concrete and steel cages were replaced with outdoor, natural landscapes. An extraordinary number of baby flamingos, gorillas, and other rare species have been born at the zoo—living proof that the new habitats have effectively replicated animal habitats in the wild. Clever innovations abound: heated rocks encourage lions to come out on cold days, elephants put on shows that end with one holding a sign in her trunk that reads *Don't Buy Ivory,* and an actor in fatigues talks about the life of a park ranger in Africa. In one year alone, the zoo drew a record nine hundred thousand visitors; city school children voted it their favorite destination for a field trip.

And, today, Willie B. leads his harem of three female gorillas through the Ford African Rain Forest, which is landscaped with plants that mimic those that grow in his native Cameroon. His daughter Kudzoo lies contently next to her mother. "Life," writes reporter Scott Allen, "is good for Willie B.—and for the renamed Zoo Atlanta."

Life, too, is now good for Keiko. Animals in captivity can thrive if and when they are treated appropriately, given the appropriate amount of space, and allowed to live as they would in the wild. Today Keiko watches a big-screen television that displays videos of Orcas in the wild;

he listens intently to the sounds they make and appears attracted to the screen. Someday, he will be released into the wild. Other creatures may not be as lucky as Keiko, but they can certainly be treated more fairly in captivity. They are not prisoners, nor is mankind their keeper. Rather, they are ambassadors of the wild—in some instances, the last ambassadors— and, as such, must be treated with the same respect and courtesy that any foreign dignitary would be given on a visit to America. Concern for a captive animal's well-being must not just come first and foremost, but must also be kindly, affectionately, and knowledgeably given for as long as that animal remains in captivity. Then, and only then, can the concept of captivity change from that of a loss of freedom to a life-long life-giving gift of care.

Some Ways to Promote More Natural Captivity

1. If you buy or use any product that has come from an animal that was raised in captivity, be sure that the animal was bred and cared for humanely. Find out by either visiting the facilities yourself, such as a local chicken farm from which you buy your eggs, or by contacting the ASPCA (American Society for the Prevention of Cruelty to Animals) or APHIS (Animal and Plant Health Inspection Service) in the locale from which your product originated.

2. Vehemently protest exhibits that focus solely on animal performance as well as petting zoos, roadside attractions, and operations that use animals to lure in customers—even if it is for a one-time occasion. More often than not, the keepers in such facilities or during such promotions have neither special training nor any particular regard for the animals in their care. However, keep in mind that many animals, especially the more intelligent ones, need novelty to keep their minds sharp and their spirits healthy. So elephant, dolphin, and whale shows do provide valuable mental exercises for such animals, particularly when the animals are always being taught new acts or are allowed to participate in what is called a "tandem creative," or a request from their trainer to "do something creative together"—an activity that so delights dolphin pairs in captivity that they have been heard to squeak with pleasure prior to working on a tandem movement together.

3. If a nearby facility has not passed the test that gives accreditation by the American Association of Zoological Parks and Aquariums (AAZPA), discover what zoo officials as well as city and state health officials plan to do. Will the facility be closed or will improvements be made to earn accreditation? If the intention is to close the zoo, find out what will happen to the

animals. If the intention is to keep the zoo open, keep after officials to make necessary improvements rather than maintain the status quo.

4. If your area offers an accredited zoo and/or aquarium and you are satisfied with the ongoing quality of its exhibits, provide financial support through membership fees, purchases at the zoo or aquarium shop, frequent visits, and tax-deductible contributions. Tell others about the facility. Become a "friend of the zoo" and offer to adopt a particular animal and pay to feed it for a month or a year. Volunteer to lead tours of the zoo or stand by exhibits to answer questions. Make a visit to the zoo or aquarium part of the "tour" you give to your out-of-town visitors.

5. Before you can ascertain the quality of each of the exhibits at your nearest facility, first understand the biology of the animals that are living in particular enclosures or habitats. After you find out what an animal needs, observe whether the enclosure provides this

need. For example, animals that like to take dust baths need dust or sand in which to roll, aquatic creatures sometimes need to rest so their exhibits need to include a place in which they can haul themselves out of the water, and some animals need rubbing posts to keep horns, claws, and hides in shape. Your own research will provide you with valuable information that you can use for your own observations and for sharing with zookeepers in order to inspire them to do a better job of providing what the animal needs.

6. If possible, gather together supporters or hold a neighborhood fundraiser in order to raise the money for special habitat needs, such as a climbing apparatus for those animals that need exercise or challenging playthings for animals that require continual stimulation.

7. Observe the animals in a facility and note whether they are spirited. A good facility will provide its animals with more than

just simple nutrition and sanitation. Stimulation can come in many forms, such as experimenting with new ways of feeding to encourage activity—putting honey inside a log to encourage bears to dig out a "hive," for instance, or giving sea otters clams and rocks so they can replicate feeding in the wild by floating on their backs and smashing open the clam with the rock. When animals are spirited, they are more likely to pounce, dive, climb, build, play, and exhibit other natural behaviors.

8. Because most zoos rely solely on grants from local governments, private donations, and ticket sales, help raise money for good facilities. Too, make sure at least some of the money obtained by zoos is going toward habitat restoration and conservation.

9. Accept that animals possess intellect and emotions. Because of this, refuse to see wild animals as "lesser" creatures that deserve or require less but, instead, as creatures with a unique and different consciousness that can benefit and enliven both the natural as well as the human world.

10. Support the lively, witty, and energetic philosophy of individuals such as Chris and Martin Kratt, who, with their successful PBS wildlife series *Kratts' Creatures,* show kids (as well as grownups) "how cool the animals are we share the planet with. We want kids to make decisions that will keep animals around." Keep abreast of happenings in the "wild kingdom" of captivity by renting videos, reading books, visiting different facilities, and participating in campaigns to return captive animals to the wild.

Environmental Showdowns: Creating Natural Alliances

"... perhaps no society other than the contemporary one has had so urgent a need to reconsider its motives in how it defines and uses nature. To untangle the knot of reasons and platitudes that binds us to our present understanding, we must consider not only the claims of the present-day interpreters but also those earlier transformations that have set the stage for the contemporary confrontation with the natural world. We must ask, in short, where 'nature' came from."
—associate professor of environmental studies at York University (Ontario, Canada) Neil Everden, from *The Social Creation of Nature*

Centuries ago hunting, trapping, fishing, and leveling acres of trees for homestead and hearth were part of a way of life for this country's early settlers. Deer may have been perceived as beautiful creatures, but they were more greatly appreciated when they were cut into bite-sized pieces and served at the dinner table. Raccoons, too, may have been credited with being clever critters, but they were more greatly coveted for their pelts, which earned money for men as well as young lads who were patient enough to train their coon hounds and capable enough to aim a gun to obtain the pelts that would help them to purchase much needed supplies for the family. Trees were magnificent, but there were far too many of them and they were in the way; to build a home and then to

Orr's Island, ME, 1958 © David Vestal 1997

stay warm in that home, the trees simply *had* to be cut down.

This country's early settlers most likely gave little thought to just how many deer or raccoons were being killed on a weekly basis or how many acres of forest were being leveled at will without regard to the creatures that were living in the trees or the integral role the trees were playing in the prevention of soil erosion. After all, game was plentiful and the trees were certainly abundant. At that time, the sheer numbers alone of all living things revealed that nature was considerably in surplus; there was, in fact, the perception of *too much of nature*—it overwhelmed mankind and, because of this, dominated mankind. There were no footpaths already carved into grand mountains or through forests, and there was nothing that protected man from being perceived by predatory creatures as just another tasty link in the food chain. Nature owned the land, and, because of this, mankind was

an intruder. And, until mankind learned over the years how to fashion better tools and more accurate, deadly weaponry that would not only protect it from nature but also help to subdue nature, tame it, and, ultimately, dominate it, mankind was at nature's mercy.

Perhaps this is why, thousands of years ago, mankind so revered nature and sought to connect with this powerful force in some way. "What is my place in all this?" early man must have asked while looking out over vast, dramatic landscapes that teemed with wildlife. Prehistoric ancestors ventured out into this world and then retreated, spending hours daubing paint on their cave walls, painstakingly depicting the creatures with which they shared the primeval landscape. Later, the ancient Egyptians, Greeks, and Romans chose to worship nature through the personification of natural phenomena in their gods and goddesses. Such deities as Neptune, Persephone, Iris, Ra, and others took on roles in myths that were created to explain nature's mysteries to the people. Medieval Christian and Buddhist monks studied the trees, birds, stars, and flowers in order to "learn" the mind of God and therefore understand mankind's "mission" on earth. During the Renaissance, poets and philosophers fell in love with nature, as did eighteenth-century poets who translated pastoral scenes into precise verse and nineteenth-century Romantics who let their imaginations run as wild and free as nature.

But when such natural sciences as biology, geology, astronomy, botany, and zoology came into vogue, rather than pray, philosophize, or wax poetic about nature, the urge was toward a more objective, detailed observation of nature. The question "What is my place in all this?" was replaced by, "How can all this be explained?" Many of nature's mysteries were examined, tested, researched, and then miraculously solved by great scientific minds; like Sherlock Holmes, such thinkers proved that the cause of many of nature's curiosities were really quite "elementary." Thousands of species of wildlife and wild growth were collected, identified, and then assigned human nomenclature, detracting from their former allure. As nature was being so painstakingly tracked and trampled, for safety reasons the concept of God was promptly elevated into the heavens rather than treasured in the hearts and souls of all living things that inhabited the earth.

Nature was, after much analysis and cataloguing, finally being put in its rightful place. Inspired by the easy understanding of nature provided by science, explorer-naturalists by the droves set out to conquer the wild spaces and the wild creatures; like travel agents intent upon booking for an upcoming holiday, they returned from their adventures with grand stories and great kills, describing the "exotic" locales in such a way that would entice others to venture forth and carve out their own piece of the wilderness, the prairie, the seacoast, or the mountains.

Nature, people were quickly finding out, was not so overwhelming after all. There was nothing that could be left standing after saws and plows had had their way with the land. A ferocious charging bear could easily be brought down with a bullet or slowly tormented to death in the jaws of a steel trap. Poisons could quickly eradicate pests. Rivers could be diverted. Tunnels and tons of dynamite could loosen riches from the land. Mountains could even be made into molehills. Nature, finally, was mankind's; now mankind could do with nature whatever it wanted to. After this, nature, like comedian Rodney Dangerfield, could "get no respect."

The Clovis people, or the first identifiable American Indian people who lived in the area now known as New Mexico from about 9500 B.C. to 9000 B.C., created a society in which its political leadership fell to the dominant male, who based his own authority to lead from his well-known and much talked about exploits as a hunter and provider. Hunting skills were passed down from generation to generation, vital not only to individual survival but, as well, to familial survival. The Clovis people faced extinction every day; America in the late Ice Age was a tough and unmerciful place in which to live, with harsh weather conditions and huge mammoths and long-horned bison that freely roamed the land. Clovis hunters competed one-on-one for food with such fierce predators and scavengers. One critical mistake, and a hunter could suffer a devastating injury. If he died, his family was immediately at the mercy of the elements and the land.

Although the Clovis hunter was both courageous and strong, he also was respectful of the power of the creatures he faced. In coming face-to-face with his prey, the Clovis hunter often addressed the animal spirit in the way taught to him by medicine men. He might have first apologized to the beast for what was to come (or what he hoped would come), explaining his needs and the needs of his family. He may have asked the animal not to be angry with him and, in turn, offered the creature an assurance that its body would be treated with respect. After slaughter, the carcass would be butchered in a special way, with some parts placed on display or disposed of ritually so that the animal's "life-force" would return to its home, regenerate its flesh, and come back another time. Although the point was not so much to assure continued abundance but to respect the sustenance provided by such valuable creatures, these early people could be perceived as the nation's first conservationists and animal rights activists. Descendents of the Clovis, the Naskapi Indians of Labrador, reveal such a consciousness in *Ati'k'wape'o, The Story of Caribou Man*: "He who obeys the requirements is given caribou, and he who disobeys is not given caribou. If he wastes much caribou he cannot be given them, because he wastes too much of his food—the good things. And now, as much as I have spoken, you will know forever how it is. For so now it is as I have said. I, indeed, am Caribou Man. So I am called."

Although the Native Americans always upheld a strong environmental consciousness and respect for the land on which they resided, the gradual decimation and devastation of America and its living things created little or no impact on mankind's consciousness until 1864, when the person credited as being America's first environmentalist sought to warn Americans of their wasteful ways. Vermont lawyer George Perkins Marsh foretold of the earth's environmental problems and developed a scientific branch to deal with them, now known as ecology, or the study of the interrelationship between all living things and their environment. In his book *Man and Nature: Physical Geography as Modified by Human Action,* he anticipated the current crisis in pollution, overpopulation, and vanishing resources, concluding that mankind was, in no way, an asset to the earth. "Even now," wrote Marsh, "we are breaking up the floor and the wainscotting and doors and window frames of our dwelling to warm our bodies and seethe our pottage."

Over one hundred years later actor, producer, and director Robert Redford, who also sits on the boards of such organizations as Friends of the Earth USA, the Environmental Defense Fund, and the Natural Resources Defense Council, wrote: "We've poisoned the air, the water, and the land. In our passion to control nature, things have gone out of control. Progress from now on has to mean something different. We can't keep using up one place and moving on to the next. We're running out of places, we're running out of resources, and we're running out of time.... Native Americans live with seven generations in mind. Our leadership right now seems to be without plans for even one generation of Americans." Even a brief rundown of some of the more devastating environmental disasters caused by mankind throughout the years is enough to depress even the most skeptical of non-environmentalists: the 1946 atomic bomb test explosion that contaminated the once Eden-like Bikini Atoll; a thick London smog that killed hundreds in the 1940s and 1950s; the thin-shelled peregrine eggs that resulted from the use of DDT in the 1950s that nearly drove the species to extinction; the first in an on-going series of big oil spills, *The Torrey Canyon,* which occurred off the Cornish coast in 1967; and fallout from the radiation of the 1986 Chernobyl disaster, which spread over Europe.

On and on goes the laundry list of mankind's dirty deeds. The vision of progress that once started out with innocent enthusiasm and steadily drove the world's economy after World War II certainly has enriched the lives of many over the short-term, but, unfortunately, has greatly impoverished Planet Earth for the long-term. The tiny, blue-capped Gurney's pitta, for example, is one of many species that now teeters on the brink of extinction as a result of destruction of lowland forests in Thailand and Burma. In South Africa, poachers still ruthlessly slaughter elephants for the price their ivory tusks will bring (leaving behind the massive carcasses and bloodied faces of these gentle creatures to rot in the sun) and rhinos for the nonexistent "medicinal" (that is, sexual) properties purportedly bestowed upon those who ingest a concoction made from its powdered horns. Worldwide, uncontrolled fishing has reduced stocks of almost all commercial species, some to less than one-tenth of their former numbers. The world's rainforests are being burned and bulldozed so fast that half of those remaining today may vanish in the next forty years. Each day fifty to one hundred species of animals and plants become extinct as the habitat of these wild, living things is gradually destroyed. The number of motor vehicles in the world is set to increase steadily by fifteen million a year until at least the year 2010; the corresponding rise in carbon dioxide emissions, lead, sulphur dioxide, and other combustion products will accelerate climate changes caused by the "greenhouse effect." Once-pristine springs that bubble from the rock near the top of the continental divide now show traces of lead from car exhaust fumes. Industrial wastes have seeped deep into the land and impregnated the soil and groundwater with chlorides, sulphates, heavy metals, and other chemical pollutants. Global warming is a cause for daily concern for everyone, but, as well, for the people of Venice, Italy, who know that continually rising sea levels threaten their homes and the future of their city.

On and on the tragic consequences of mankind's actions against the environment goes; upward and upward the ecological price tag soars. The richness of bank accounts has, over the years, compromised the richness of life; the environment cannot continue to pay such a high price for what mankind calls "progress." The intellectual revolution that revised the view of nature from that of a caring, living, breathing, nurturing Mother Earth to something with less value and transcendental purpose—to something more inert, to something more like a machine than a living organism, stripped of its sacred values, its metaphysical meaning, its diverse harmonies, and its powerful relationship with mankind—has convinced mankind that there never was a vital web that connected human societies and cultures with the natural world, even though mankind has, for thousands of years, coexisted and coevolved with the natural world. "We have inflicted awful wounds on the Earth," writes Jonathon Porritt, Director of Friends of the Earth UK and author of *Save the Earth*, "and are now caught in the trap of trying to heal these wounds by prescribing more of the same Earth-defying remedies. In the process, the human spirit has also come under constant attack. Many of the intangible values (a sense of community, a pride in serving others, a love of the land and the rhythms of nature, spiritual enrichment) that once provided comfort, fulfillment, and meaning are increasingly denied to people, written off as so many wisps of nostalgic romanticism. It is not just the Earth that has paid the price of our obsessive pursuit of industrial progress, but that fragile part of us that responds to a higher reality than material wealth."

Mount Desert, ME, 1958 © David Vestal 1997

Today you probably know more about the surface of the moon than you do about the biological communities outside your back door or in your front yard. Such ignorance, however, can be deadly. You, as well as all of mankind, *must* realize how entirely dependent you are upon the plants, animals, fungi, and micro-organisms that share the world with you. Without such things, you will eventually die. You need nature although, in truth, nature does not need you. Nature can feed itself, protect itself, reproduce itself, and thrive by itself for an eternity. It does these things and, as well, feeds mankind, provides many of the drugs and other products on which the quality of your life increasingly depends, and offers the promise of a sustainable, productive future for as long as mankind acts responsibly.

Just as your predecessors questioned, "What is my place in all this?" so too must you begin to question your place in the natural world. As well, you need to begin to consider what you feel is your responsibility to the planet, whether animals are here to serve you or you are here to respect and protect them, and whether industrial and technological progress has gone too far or not far enough. You need to consider such things because how you resolve them in your mind will determine not only how you will choose to live your life from this day on, but, as well, what your vision will be of the future of Planet Earth.

Once, mankind was powerless to change nature. Once, the world was rich in natural resources, rich beyond even the wildest of dreams. Once, clear streams were filled with fish, the world was thick with primeval forests, fields of grass stretched as far as the eye could see, and the earth was bursting with minerals like an empress's box of jewels. Once, there was no doubt that such bounties would last forever. *Once. Once upon a time.*

Now, however, after thousands of years of actions mankind has taken against nature, every continued action has an even greater impact and every new action takes on a more profound meaning. Everything you or anyone else chooses to do reveals a great deal more of the fragile division that exists between nature and mankind. Even though there is much land that has been set aside as undeveloped, even though many species are thriving or making comebacks, even though there is a greater exploration and use of alternative energy sources, and even though recycling is no longer perceived as a drug-induced, radical-'60s concept, nature, as a whole, is on life support. It cannot continue to live without your care and attention; it cannot begin to breathe again on its own, to sustain itself, and to live and grow without your help. Nature needs more allies, for it still has far too many enemies. In most parts of Africa, for example, the leopard is a highly endangered species. Yet there are still people in the world who would prefer to have the leopard draped dead over their shoulders than alive in the wild. Hunting, fishing, and trapping today provide a living for precious few Americans but are now popular recreational sports that have, for the most part, produced citizen backlash over inhumane methods of killing or provoked criticism at infrequent but moronic incidences of shooting not just at game but at *anything* that moves in or near the woods—from dogs to cows to fellow hunters to innocents hanging up their laundry in their own back yards. To many, hunters are perceived as drunken Elmer Fudds or, worst yet, inhumane murderers not just by animal rights advocates but by society in general; anglers are viewed as harmless but useless retirees who have nothing better to do with their time than to impale worms and hook no-brainer fish. Just as nature is slowly coming apart, so, too, is the fabric of society being ripped to shreds over what the right thing to do is. Environmentalists are being criticized for focusing so much on the negative; those who continue to decimate the water, the woods, and the wildlife are accused of not focusing at all on the outcome of their actions. Saving the planet has, in many ways, divided its inhabitants rather than united them.

The healing of the earth must involve the healing of the human spirit as well. Because of this, any healing action presents a powerful opportunity to converge human needs and the needs of the rest of life on earth. While you, as one person, can affect positive change on a small scale, much more is needed. The destiny of mankind and nature lies in the alliances that need to be forged between larger groups and organizations that have divergent philosophies and beliefs, that offer differing lifestyles and goals, and that have, thus far, thrived on their adversarial positions and their opposition to one

another. The economist needs to see and measure the fiscal state of the economy as well as the damage wrought by strict adherence to economic growth. The politician needs to protect the interests of the people he or she represents as well as the nonvoting constituents that share his or her district—in its forests, wetlands, deserts, mountains, oceans, lakes, ponds, streams, in the air, and on the ground. The scientist needs to provide mankind with facts and figures of ecological decline as well as viable solutions for preventing continued decline. The businessperson needs to contribute valuable products and services to the marketplace as well as ensure that the way the products and services are manufactured, marketed, and disposed of protects rather than degrades the environment.

What this means is that instead of taking pot-shots at one another, instead of pointing fingers and assessing blame, instead of suggesting what ought to be done or can be done rather than take effective action, and instead of resorting to protests, boycotts, and letter-writing campaigns that are designed to raise issues rather than affect positive change, opposing parties must begin *now* to work together to save the earth—*to forge lasting alliances that are not created in order to change mind-sets, stop production, or make "converts" out of anyone but are designed to build up an invincible force that will protect and preserve the environment without requiring that either party abandon or severely restrict or compromise its own "cause."* The logger *must*, therefore, learn how to hold hands with the conservationist in order to ensure future forest growth and harvest. The car manufacturer *must* consider prototypes from the electric-car inventor in order to reduce harmful products that are being spewed into the atmosphere and to lessen dependence upon finite natural resources. The utility company *must* welcome solar-driven and wind-driven energy products in order to help Americans learn how to thoughtfully conserve rather than thoughtlessly consume. The developer *must* listen to the person who wishes to preserve open space and seek to build more creatively as well as more responsibly. The mink farmer *must* pay attention to the advice of the humane society worker in order to understand the needs of not just the animals being held in captivity, but all wildlife. The research lab *must* learn how to perform tests without constantly relying upon live, caged subjects in order to discover that herbs and other wild growth offer inexpensive and potentially more beneficial product alternatives. And so on. The alliances that can be formed are endless; none today, however, has the potential to be as politically powerful and as naturally enhancing and beneficial as the alliance that is starting to be formed and strengthened between hunters and anglers and their long-standing adversaries, the environmentalists.

More than fifty million Americans fish and fifteen million hunt, but few environmentalists, conservationists, and animal rights protectors realize that these sportspeople have had a long history of protecting and restoring fish, wildlife, and habitat. America's 92-million-acre national wildlife refuge was started by former president and avid hunter Theodore Roosevelt. This refuge has been financially protected since 1934 because of the actions taken by Pulitzer prizewinning political cartoonist of *The Des Moines Register*—and hunter—J. N. "Ding" Darling who, along with his fellow waterfowl hunters, pushed a law through Congress that required all duck and goose hunters to purchase a federal permit in the form of a stamp. Ever since then, duck-stamp money has gone on to purchase national wildlife refuges. Others have since followed Darling's lead. To raise money for wildlife management, for instance, hunters and anglers have successfully lobbied for excise taxes to be placed on their own fishing tackle and ammunition and have begun to push for new excise taxes on an even wider range of outdoor products used by them, such as backpacks, tents, birdseed, and field guides, which they anticipate will raise another $350 million for ecosystem management. These same sportspeople have also, since the beginning of the twentieth century, *saved* game—and many species now classified as nongame—from commercial market hunting. They have helped to restore the American bald eagle, wild turkey, peregrine falcon, and Atlantic salmon. They have formed a broad coalition of thirteen Atlantic states to hammer together a management agreement that brought the striped bass from its drastic state of decline in the early 1980s to a resoundingly successful restoration. They have adopted wildlife management measures that ensure the survival and well-being of whitetail deer, black bear, and moose.

And these gun-toting, rod-n-reeling sportspeople have not wearied or wavered in their efforts to protect and restore nature. The controversial aerial wolf hunting practice in

Alaska, for example, which many thought was linked to hunters, was actually proposed by the Alaska Department of Fish and Game, led by director David "Machinegun" Kelleyhouse, who had once tried to requisition a fully automatic weapon for "wolf management." The proposed hunting practice was supported by then-Governor Wally Hickel, who explained to journalists at a "wolf summit" that "you can't just let nature run wild." But a statewide poll revealed that only thirty-six percent of hunters were actually in favor of aerial hunting. In fact, in order to strengthen their opposition against the practice, Alaskan hunters, who not only admire wolves but appreciate the vital role they play in the management of ecosystems, joined with environmentalists in forming the Wolf Management Reform Coalition to try to ban aerial wolf hunting permanently in the state.

When hunters, anglers, and environmentalists join together, their efforts not only accomplish a great deal but also benefit one another. Together, the Rocky Mountain Elk Foundation, run by elk hunters, and the National Fish and Wildlife Foundation, run by a former National Audubon Society lobbyist, have protected or restored 1.8 million acres north of Yellowstone National Park. Trout Unlimited, run by sportspeople as well as environmentalists, is on the forefront in protecting endangered species and promoting grazing reform, mining reform, hydroelectric relicensing, clean water, and forest practices. "Big green" groups such as the National Audubon Society and the Sierra Club have never opposed hunting and, in fact, recognize that the sport is a legitimate and necessary part of wildlife management, even though certain state chapters have positioned themselves, through their actions, as anti-hunting. The Natural Resource Summit of America, a political alliance formed in 1995 to save the Clean Water Act, strives for solidarity on such fronts as environmental law and natural-resources and public-lands policy. "The third meeting of the summit," reports Ted Williams in *Sierra,* "was attended by such diverse groups as the Sierra Club, the American Fisheries Society, the Izaak Walton League, the American Sportfishing Association, The Wilderness Society, the International Association of Fish and Wildlife Agencies, and the Environmental Defense Fund." And, when Senator Pete Domenici (R-N.M.) introduced a grazing bill that predominantly reserved public lands in the West for the ranching industry, opposition to the bill ranged from an alliance formed from environmental groups such as the Sierra Club and Defenders of Wildlife to California Bowman Hunters and State Archery Association and Sportsmen's Council of Central California.

"Environmentalists don't reach out to sportsmen," comments Chris Potholm, a professor of government and legal studies at Bowdoin College in Maine. "If they did, they'd be invincible. Whenever sportsmen combine with environmentalists, you have sixty to seventy percent of the population, an absolutely irresistible coalition." Too, such an alliance would provide an increasingly valuable ally to nature that, ultimately, would benefit all of mankind.

Concern for the environment is no longer the preserve of a small, eccentric minority. It cuts across the political spectrum, left to right. It unites those who want to save up as well as those who want to use up. It overcomes conflicts between policy makers, economists, and ecologists. And it speaks up for those causes and people it directly supports as well as those that may be potentially or indirectly impacted. Take, for example, the actions initiated by the American Lung Association. Fed up with reports that revealed increasing millions of Americans suffering from pollution-related breathing disorders, including asthma and bronchitis, despite passage of the Clean-Air Act, the Association sued the

Environmental Protection Agency in 1993 to make it re-evaluate its standards for particulates (fine dust particles) in the atmosphere. Almost three years later—and two days before a court-ordered deadline—the EPA responded to the persistent pressure. In a decision that EPA chief Carol Browner calls "one of the most important of her career," the agency proposed tough new standards on both particulates and ground-level ozone, a major component of smog. Such standards will, according to the EPA, *save twenty thousand lives a year* and, too, drastically reduce the amount of pollutants being released into an already greatly degraded atmosphere. Although the stage has been set for what has been called "the biggest environmental battle of the decade"— more than five hundred businesses and local government groups are lobbying *against* the rules simply because of the financial impact it would have upon them—the rules would greatly benefit the environ-ment and the quality of air that supports all living things by requiring Americans to change the way they live. Even though the unlikely alliance that was forged between the American Lung Association and the EPA started out as a battle, it ended up in a partnership in which both participants contributed their time and effort to a worthy and environmentally life-saving cause—the future of mankind and the Planet.

Alliances, in many ways, encourage the lion to lay down with the lamb. But they are vital to the future of the universe and for the creation of a better way of life. For nature cannot be restored, pre-served, or protected by any one person, any one decision, any one rule or regulation, or any one group for a long period of time. Nature can, however, be saved by many individual choices and voices allied with one another and work-ing together in one great effort to make a better world for everyone and every living thing.

Some Ways to Create Natural Alliances in Your Life

1. Pledge to do the following: "In my effort to help heal the earth, I pledge that in my personal behavior as a responsible citizen and voter I will do my best to make the appropriate decisions that will make the earth a secure and hospitable place for future generations." Ask others you know to make the same pledge.

2. Strive to use twenty percent less gas and/or electricity during the next year; encourage government officials to do the same in their offices and homes. Try to cut your car mileage by twenty percent during the next year; walk, bicycle, or use public transportation whenever possible, and encourage coworkers to do the same. Plant ten trees in the next year and look after them until they can survive without your assistance; donate ten trees to your town to plant and tend. Increase the amount of time you give to environmental as well as developmental organizations in your community;

seek to link the work of both organizations whenever possible.

3. Look into the make-up of your local and state Fisheries and Wildlife Board. Support requirements that a majority of the board members have hunting and fishing licenses. Remember, hunters and anglers are not hostile to wildlife, but *devoted* to it; they would rather a deer die quickly than slowly starve to death. Make sure, however, that hunting and fishing interests as well as wildlife interests are being promoted and protected equally. Find out where board members stand on inhumane, deplorable practices such as baiting and hounding bears, using leghold traps, or same-day airborne hunting of wolves, lynx, foxes, and wolverines as well as the use of lobster gear, gillnets, and other so-called fixed fishing gear that have been known to harm endangered whales.

4. Organize crusades to save special places in your area. Show local businesses and citizens the wildlife and beauty of such places by

taking them on tours of the land. Convince them of the vital need to preserve such sanctuaries not just for the good of nature, but also for the good of the community. Make the conservation area one that is not only nature-friendly—a place in which suburban outcasts such as fox, opossum, raccoons, and skunks can live free—but also people-friendly. Ally corporate sponsors with citizens to create well-marked trails and retired carpenters with local scouting groups to build benches, birdhouses, and foot bridges over streams.

5. Encourage lawyers you know to take up animal rights defense so they can speak up for animal rights. Today just a half-dozen independent, private attorneys make a living in animal rights law, but the animal causes they have defended have been worthy. For example, when the New England Aquarium sought to transfer two dolphins to the Navy, Boston-area attorney Steven M. Wise blocked their possible use in military exercises that would have meant certain, senseless death. Because airlines are not known for their humane transportation of animals—it is not uncommon for beloved pets to end up literally baked, melted, suffocated, or frozen to death in a plane's baggage section—pet owners need such legal advocates to safeguard animal rights. "There's no reason to limit fundamental legal rights to human beings," proclaims Wise, who, with his wife, has handled over one hundred and fifty dog cases that involve a Massachusetts law that lets towns execute or banish a dog deemed a nuisance "by reason of vicious disposition or *excessive barking.*"

6. Recognize that even though there are a great number of national parks, federal budget cutbacks are forcing them to forgo repairs on everything from historic structures to bathrooms to potholes. As well, campgrounds and education programs are being shut down or scaled back; preservation and trail maintenance projects—vitally needed to keep rare vegetation alive and trails

intact—are being postponed or scrapped. Also, the number of rangers is being reduced, limiting assistance and security to visitors and reducing protection of the grounds. Ally state and local government funds as well as private money to ensure parks remain open and adequately maintained and that park rangers stay employed. Too, encourage parks to adopt business practices and modern principles of management that will enable them to keep up with their record numbers of visitors while showcasing their treasured park with pride.

7. Encourage automakers to rethink every part of the "car culture"—from fuel to roads to engine design—in order to create a less environmentally depleting concept vehicle for the average citizen as well as for public transportation programs.

8. Before "green fatigue" sets in—too much talk and not enough action, too much doom and gloom and not enough hope, and too many unresolved problems—recognize that progress must come through a willingness to take action. Put the pressure on politicians to accelerate the pace of change. Tell them that if they refuse to protect the environment, they will not be elected or reelected. Continue to work at the grassroots level, but seek out greater media attention. Voice concerns! Demand action!

9. You need nature more than nature needs you. Starting today, learn to live economically, emotionally, and spiritually in ways that put the environment first and profits last.

10. Recognize that whether you see a forest as a mythical realm or a stock of unused lumber or whether you prefer to observe a duck or eat a duck, it is nature nonetheless. Your view is significant, but how you ally this view with an opposing view will have the greatest impact. When you can see nature from a multitude of viewpoints, you can then truly see that nature is a social creation that connects you not only with nature but with all of mankind.

The Call of the Wild: Becoming Intimate with Nature

"You can grow intimate with almost any living thing, transfer it to your own emotion of tenderness, nostalgia, regret, so that often of a relationship one remembers the scene with the most affection. A particular line of hedge in a Midland county, a drift of leaves in a particular wood: it is only human to imagine that we receive back from these the feeling someone left with them."
—English writer Graham Greene

"Weasel! I'd never seen one wild before.... Our eyes locked, and someone threw away the key.

"Our look was as if two lovers...met unexpectedly on an overgrown path when each had been thinking of something else: a

clearing blow to the gut.... It emptied our lungs. It felled the forest, moved the fields, and drained the pond; the world dismantled and tumbled into the black hole of eyes. If you and I looked at each other that way, our skulls would split and drop to our shoulders. But we don't. We keep our skulls. So."
—Pulitzer Prize winning writer and naturalist Annie Dillard, from Teaching a Stone to Talk

For years, Mrs. Susan Gilbert has helped her first and second graders in Lexington, Massachusetts, raise butterflies and then release them into a nearby field for their flight to Mexico—not just one butterfly a year from the single caterpillar the science department buys every elementary school class to teach students about metamorphosis, but *all* the caterpillars she finds. "I don't know if I started looking for them, or they started looking for me," Gilbert comments about the numbers of caterpillars she collects from the hood of her car, the playground, and everywhere else she goes. "But somehow, it became a thing." Her enthusiasm for the development of the butterflies—she even once took a caterpillar with her into the hospital during scheduled surgery so she would not miss a crucial stage in its metamorphosis—is shared by her students. Each school year, they watch over their brood like expectant parents, marveling as the eggs hatch into tiny caterpillars that constantly eat and grow until they spin themselves into a chrysalis and then, finally, break out as butterflies—winged and colorful.

Over the years the process of hatching, raising, and releasing the butterflies has gone smoothly.

But then, in 1996, Mrs. Gilbert got an unexpected gift for her butterfly-appreciating students. A breeder in Westford gave her a cluster of monarch butterfly eggs he was going to throw out. He felt the butterflies would not survive because they had hatched late in the season—too late to survive their migratory flight to Mexico from New England because the upcoming cold weather would mean fewer flowers for them to feed on as they journeyed south. The students gladly "parented" the butterflies and then, as cold weather set in, enthusiastically launched a campaign to get the butterflies safely south so they could survive the winter and then return to their "native" north to lay eggs and die. Even though scheme after scheme seemed to always fall through—someone's grandmother was supposedly coming to visit from Florida and might be able to take them home with her, a student whose family was going to Disney World might

Corea, ME, 1958 © David Vestal 1997

be able to bring the butterflies as well, another student's father was scheduled to go on a business trip to Texas that was cancelled at the last minute—the students remained optimistic. "We knew they would have died if they flew on their own," explains student Kate Hoffmire, who is six years old. "It would be like raising a baby, then letting it die."

Then, in mid-November, the students' unrelenting optimism paid off. Southwest Airlines, flying out of Providence, Rhode Island, donated a seat for Mrs. Gilbert to fly to Texas; she could carry the seven monarch butterflies in a box covered by netting, then release them in the Lone Star State. "Are they flying Butterfly Air?" joked six-year-old Chloe McGuffin when Mrs. Gilbert broke the good news to her class. "I've never heard anything so silly," sniffed Anna Lubroff, also six, although she freely and proudly admits to spending her free time hunting for flowers to feed the butterflies, growing milkweed for them, and, when she could no longer find wildflowers, preparing mashed-up bananas to feed them or putting drops of water on her fingers so the butterflies could drink.

"We're happy to help out the butterflies," proclaims Kristie Kerr, a spokeswomen for Southwest Airlines. "We actually got another request from a butterfly breeder in Cleveland." Six-year-old Emily Fenn is ecstatic. "We've had them since they were baby eggs," she explains. "We got to be good friends with them."

To be so passionate about one of nature's living things is what connects the human spirit with the natural spirit. It is what brings you in harmony with nature; it is what makes you feel that you are a part of all that you see, smell, touch, taste, and hear in the world around you—a world you share with nature. When you can experience such a passion for nature, you can then even *know* nature; you can know the eagle, the fox, the carp, and the centipede in ways that go beyond textbook descriptions because you *share* in their sympathies and sensitivities, much as you would share in the sympathies and sensitivities of someone with whom you spend time and really like. You can intuit the impact a single bee can make upon a single blossom because you *know* the bee and you *know* the blossom. You can develop a conscious awareness of what it means for sap to stir within a tree and how wonderous it is to stretch roots deep into the living earth to feel the tickle of a passing earthworm or the caress of luscious rainwater as it envelops you. You can become one with the weather; it no longer chills you or soaks you or heats you or fans you but *is* you.

When you can become so intimate with nature, you become nature and, in turn, nature becomes you. Such a passion for nature is not like a lover's wild embrace—both terrifying and enticing—but like the feel of a lifelong partner's arms wrapped around you—intimate, deep, pure, lasting, uniting, true, safe, secure. Such a passion is a fusion of the mind—the intellect and the intelligence—and the senses. It is a joining of the heart and soul with the heart and soul of something outside you that is so complete that you are unaware of any division between you and this something else; you are a complimentary blend without egos. Chinese sage Chuang Tzu once described this passionate, intimate union between mankind and nature when he said, "I do not know whether I was then a man dreaming I was a butterfly, or whether I am now a butterfly dreaming I am a man." The two disparate living things, to him, were both one and the same.

With this passion, however, comes a grand responsibility: you *must* protect the vulnerability created by such intimacy. For in becoming so intimate with nature, you have, as well, made the decision to live not just for yourself but, as well, for nature. So all your thoughts, actions, feelings, desires, dreams, and goals must afford protection and security toward nature as well as express your undying love and passion for it.

There is a flip side, a truly dark side, to having such a passionate love of nature. Like the love between two people that can sometimes take a wrong turn and end up in an obsessive, destructive version of what was once a pleasurable and meaningful emotion, so, too, can a simple, passionate love of nature turn into a detrimental, damaging, and even deadly fixation.

Take, for example, the case of Thomas Kral, lepidopterist. His dazzling butterfly collection rivals that of even the finest museum collections. Where the National Museum of Natural History, the American Museum of Natural History, and the California Academy of Science *combined* have a total of twenty-eight specimens of the rare Uncompahgre fritillary, Kral's collection boasts a phenomenal nineteen specimens. But his collection goes way beyond the boundaries of a harmless hobby; it contains 1,637 illegal butterflies, including members of eleven out of sixteen species of North American butterflies currently protected under the Endangered Species Act, along with two threatened species and numerous butterflies afforded protection in national parks, national forests, and national wildlife refuges.

And yet this represents only *two percent of his entire collection*. To this day, authorities who prosecuted Kral are stunned. They wonder, how could Kral, who professed to love such beautiful and delicate creatures, collect them so rabidly and compulsively that he was *decimating* certain species of the population, nearly collecting them to extinction, until all that would remain of a particular type of butterfly would be a lifeless corpse entombed under glass to passively peruse rather than actively experience?

Kral started collecting butterflies after he found his first polyphemus moth at the age of six. In the years that followed, he devoted an extraordinary amount of time and energy to his passion. "I collected butterflies in my backyard," he says about his early collecting years. "It was the joy of finding something I'd never seen before.... Aesthetically, I like looking at butterflies; I find them beautiful." Each butterfly was then neatly labeled with a miniscule tag and then mounted with almost surgical precision. Every butterfly collected by Kral had its own story—where it had been caught and when. He often went out on collecting field trips with twenty thousand envelopes and would sometimes fill up to *nine thousand per trip,* an amount that astounds even the most dedicated lepidopterists. But many renowned collectors are also well aware of the fine line that exists between collecting for the sheer joy and pleasure it can

bring and the desperate, controlling obsessive drive to amass that can be overwhelming. "With butterflies, you can take them home, you can *possess* them," explains a premier collector who prefers to remain anonymous. "The purchase of a butterfly is not the same as the hunt, the long drive home, the pinning, the spreading of the wings. There is a whole process that draws the maniacal collector in. You cannot separate the pure love of collecting from the desire to obtain."

When U. S. Fish & Wildlife Service officials closed in on Kral and his collection in late 1993, what they discovered was a hobby that had turned into an illegal and tragic obsession. There was evidence that Kral had participated in a decade of trade in federally protected species with a *modus operandi* that was likened to that of a serial killer. As well, preservation of species was done in such a way that senselessly tormented the fragile creatures. Chrysalises were stored in a dark shed so when the new butterflies emerged, they would not see the light. The darkness kept them still and prevented them from moving around, which might potentially damage their wings. Once their wings had become pumped with blood and the butterflies were ready for flight, they were immediately placed in glassine envelopes and stored in a refrigerator to make them easier to preserve. Hundreds of butterflies were discovered in this way; Fish & Wildlife officials who searched Kral's home and opened his refrigerator door saw hundreds of beautiful butterflies slowly moving, slowly dying in their entrapments.

Kral was well aware that what he was doing was illegal; he knew that he was poaching and trafficking in butterflies that were protected by the Endangered Species Act as well as by state, federal, and Mexican wildlife regulations. But even though he had been caught once collecting in Florida's Everglades National Park, he pleaded ignorance of the law and got away; he continued to add to his collection through illegal methods. "Myself, I pretend to be a birdwatcher when collecting adults on the wing," Kral wrote in a letter to another butterfly poacher, "quickly stashing my net & using binoculars when someone approaches. Also have worked out a scheme to elude authorities. I will just pull out a book on Western plants & say I am a student identifying plants in the wild. . . ."

On August 1, 1995, Kral was sentenced to three hundred hours of community service, three thousand dollars in fines, and three years' probation. His recommended community service was to consist of "educational activities, directed toward informing society about the National Wildlife Protection Laws." For Kral, his passion for butterflies had tragically turned into a crime of passion, one in which his need to possess was so strong that, in reality, he "loved" his beloved creatures to death.

To those who seek to collect from nature in some way, there must be a conscious consideration paid to exactly what constitutes a passion that is based upon good intentions and positive, rewarding experiences and a passion that is forged by compulsive actions and may have life-threatening outcomes. The sport of mountain climbing provides prime examples of both passions—one that is pleasurable and exhilarating; one that is draining and deadly. No one sojourn up a majestic peak has ever put these two passions into clearer perspective than the ill-fated May 1996 climb of Mount Everest, which resulted in the deaths of eight people—including two of the world's most respected mountaineers and guides, Rob Hall and Scott Fischer—as well as physically, pyschologically, and spiritually scarred more than thirty other participants, some of whom actually reached the summit and made it safely back to base camp.

What had started out, for the Everest guides, as personal passions for mountain climbing years previous had taken a different turn somewhere along the way when, instead, the passion was transformed into the formation of competing business ventures that pressured each guide, in his own Everest-scaling adventure package, into taking as many "paying customers" as possible to the summit for a "peak experience." The more "summit-satisfied" customers they each had, the bigger their bragging rights became for their individual businesses so they could gear up new clients and entice returning customers to the mountain. With between sixty thousand and one hundred thousand dollars, clothing designed to protect the body to eighty degrees below zero, portable oxygen tanks, high-altitude equipment, satellite phones, and a multitude of other high-tech gear—along with the guides' own enthusiasm—it seemed that anyone could "lay claim" to Everest. In fact, when socialite-alpinist Sandy Hill Pittman, who was led down to safety during the tragic climb, described her ascent with Fischer's group on the Internet, one cyber correspondent asked her whether there were ". . . any permanent markers at the summit. Flags, or plaques, or anything like that? A gift shop, perhaps?" Pittman omitted describing the most enduring mementos that adorn Everest's higher reaches—the bodies of dead climbers.

The spring 1996 death-storm on Everest posed, for the very first time, a question that strived to put into perspective the passion for mountain climbing. Was it a sport—something to be enjoyed—or was it something that could be bought? Was the mountain something to be treated with respect, or was it something to be conquered, possessed, won? Jeff Blumenfeld, editor and publisher of *Expedition News*, summed up

the Everest experience and seemed to offer some answers when he wrote, "You can be hooked up to a Website, you can call anyone on a sat phone, you can have the latest high-tech gear, and the mountain can still win."

But if your goal is to win or lose at any cost—even if the loss is of your own life—then is what you are doing out of love, or have you crossed that fine line between passionate love and uncontrollable desire? A less dangerous but still demanding mountain climbing sport involves a collecting experience called "peakbagging." Organized as a sport in America in the 1940s, it involves not just visiting a mountain peak but "collecting" it—adding it to the list of mountains that have previously been scaled and, thus, conquered. Some peakbaggers are members of the Four Thousand Footer Club, which involves climbing the forty-eight White Mountains that are over four thousand feet high. The Adirondack Forty-Sixers climb, or "collect," forty-six peaks in their area. The New England Hundred Highest, which includes peaks from the 6,288-foot Mt. Washington to Maine's "tiny" 3,764-foot Mt. Coe, contains several mountains that are trailless; two require permission from landowners in order to climb. The Northeast 111 includes the traditional forty-six Adirondack mountains and sixty-three New England 4,000-footers. The Catskill 3,500 Club has thirty-five summits over thirty-five hundred feet. The South Beyond 6,000 Club has thirty-nine peaks. The Western version of a "Grand Slam" involves scaling all fifty-four mountains in Colorado that are over fourteen thousand feet.

As a peakbagger, you can stay in one area and add to your peakbagging experience by attempting repeated climbs of the same mountain during the winter or striving to set personal records for the fastest climb—or even the slowest. You can peakbag across America, setting your sights on scaling the highest peak in every state. Options and opportunities for peakbaggers abound; the lists, clubs, groups, and shoulder patches that can potentially be earned go on and on. And, because most climbs can be accomplished in a day—including time taken to enjoy a gourmet picnic lunch on the summit as well as picture-taking sessions—with minimal equipment and by paying just a nominal fee, peakbagging can be a delightful experience for those who simply enjoy a day hike in the woods.

Most peakbaggers like to climb the same mountains over and over again. They enjoy the experience of getting out in different seasons to some of their favorite peaks, explore a variety of trails—sometimes even blaze their own—strive to reach the summit at different times during the day in order to watch the sun rise or set, climb in all kinds of weather just for the "experience," and seek to perfect the "perfect" peakbagging meal. Bruce Scofield, in his book *High Peaks of the*

Northeast, describes the peakbagging experience as one in which "... the activity is just too satisfying to give up—the process has become the goal." When John Gallagher of Hanover, New Hampshire, was nine years old, he set a goal to complete the White Mountain Four Thousand Footers before he was thirteen. He finished at the age of twelve. "I was tired but I was really happy," he says of his experience. "But I was also a little sad, because I was done, and it was something really fun to work toward." While some challenge-oriented peakbaggers have died in their attempts to complete their lists or have explored bold, new ways of peakbagging, such as seeing how many times a list can be repeated or climbing at night or, for at least one

Willow Creek, CA, 1959 © David Vestal 1997

peakbagging die-hard, peakbagging while naked, most peakbaggers would agree that obsession is not the central problem in peakbagging. Instead, it is having to reach the summit in order to "bag" it. That means a good pair of hiking shoes, good physical conditioning, good food, and a good attitude; turning back does not represent as terrible a loss or sense of failure as it would in not reaching the summit of Everest. To the peakbagger, the mountain will always be there, and tomorrow is a new day.

"To those who have struggled with them," Sir Francis Youngblood has written, "the mountains reveal beauties they will not disclose to those who make no effort. That is the reward the mountains give to effort." To New Hampshirite David Hooke, each new peak he scales is a "'mental tchotchke'—a knickknack of the imagination." His last mountain "keepsake" was Mt. Tecumseh. "It hadn't been described in the guides as having a particularly memorable summit," he remarked. "But I got to see Moosilauke and Washington from an angle I had never seen before. Oh boy. It was worth it." That anyone would even *want* to try to climb a mountain represents the ultimate faith one has not just in one's self but, as well, in the world. For while the actual climb up to a certain height, in reality, means little, the *choice* to take such a high road at all—one that offers challenge as well as reward, escape as well as security, risk as well as triumph—is what the heartfelt desire to climb a mountain is all about. There is a whole new world at the top of a mountain, a world that can never be experienced until the ascent is made. And yet the ascent must be made not just from a strong will, strong legs, and strong lungs, but as well, from a deep love. As mountaineer George Leigh Mallory once remarked, "What we get from this adventure is just sheer joy. And joy is, after all, the end of life."

To those first and second graders in Mrs. Gilbert's class, the yearly classroom metamorphosis involves more than just the transformation of their caterpillars into butterflies—into beautiful, wonderful, flighty creatures. It is, as well, a metamorphosis of themselves, in which each child gets to experience a sense of wonder and pleasure from the living world and learns that nature can be loved and cared for with the heart as well as the head. "One vivid memory remains of passing through the city," American nature writer Edwin Way Teale once detailed. "A small boy, five or six at most, had picked up a dead monarch butterfly from a pile of litter beside the street. He was standing entranced, bending forward, oblivious to all around him. It seemed as though I were looking at myself when young. A door was opening for him, a door beyond which lay all the beauty and mystery of nature."

To love nature, you must first answer the call of the wild. Go beyond merely recognizing that nature exists. Adore nature, revere nature, respect nature. Become intimate with nature. Whenever you do, both you and nature will grow and benefit from the experience.

Some Ways to Become Intimate with Nature

1. If you can, create a healthy environment for animals that have been "cast-off" from a former way of life. Adopt a greyhound, for instance, that has been retired from racing life. Or provide financial support for those who have taken on the responsibility of caring for much larger animals. The nonprofit Elephant Sanctuary, just outside Hohenwald, Tennessee, is a 112-acre farm run by Carol Buckley and Scott Blais for elephants that were formerly in dire straits. Tarra, a twenty-two-year-old elephant who has been with Buckley since the creature was six years old, had been imported from Thailand by a tire store owner who used the elephant to promote his business; she was inhumanely confined in a truck until Buckley took out a twenty-five thousand dollar loan and purchased the elephant. Barbara, an emaciated and scarred twenty-six-year-old former circus elephant, and Jenny, a twenty-seven-year-old member of a small traveling circus permanently hobbled by a leg injury that was left untreated, now reside in the first natural-environment elephant sanctuary in the United States. Taking care of these Asian elephants is an expensive, full-time job; the Sanctuary is supported by donations.

2. Support zoos and wildlife sanctuaries in which the directors live on or close to the premises. Living at the zoo on a full-time basis provides greater security, symbolizes the depth of commitment to the animals, ensures visiting researchers a place to stay that is near the animals they wish to study, and creates an intimate link between humans and animals.

3. Transform abandoned, rubble-strewn lots into gardens of hope. Plant annual and perennial flowers, herbs, and vegetables. In so doing, you produce beautiful and nourishing harvests as well as attract songbirds, butterflies, and honeybees.

Too, you create a hands-on intimacy with nature in inner-city areas that others might not think would support anything other than poverty, violence, and indifference.

4. Participate in more "exotic" ways of getting outdoors than the usual outdoor sports. Rather than swim laps in an indoor pool in the summer, swim in a nearby lake. Vary your running route to include off-road as well as on-road options. Instead of running in a 10K road race, for example, run through the woods or along the seashore—or even up a mountain! The Mount Washington Road Race involves a 7.6 mile race to the summit; the Fila Sky Marathon requires runners to run up to fourteen thousand feet, and then back down to the starting line.

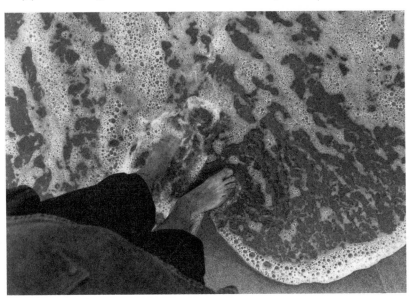

Cape Cod, MA, 1954 © David Vestal 1997

5. In the summer, pick berries in the wild. Year-round, gather herbs. If you are new to foraging, make sure you know what the native berries look like and when and how herbs should best be picked. Do not eat or make an infusion from anything you have not first researched.

6. Support and/or participate in school programs that send children out into the fields on actual "field trips." Provide seeds and seedling containers for children to start plants in their classroom. Work with parents and teachers in constructing window boxes, flower beds, and garden plots in which the children can plant what they have grown.

7. Become a nature watcher. Learn how to spot wildlife. Observe living, growing things from season to season and year to year; note any changes. Get outside in all kinds of weather; see how nature reacts to the weather.

8. Spend an hour outdoors every day. Set your alarm clock for an early morning hour and take a walk around your neighborhood at that time. (If desired, bring along a companion for safety.) Take a nature walk at a nearby bird or wildlife sanctuary. Pick a favorite outdoor spot and go there often; become intimate with that spot by observing it at different times during the day.

9. Create something that is made totally out of items you gather from nature. Make a seasonal wreath without wire. Dry aromatic herbs for a potpourri. Build a sandcastle on the beach. String cranberries and popcorn to decorate the Christmas tree. Make a rock garden from rocks in your yard. Bake a pie from berries you collect in the wild. Create a "nature-mobile" from sticks, feathers, shells, and vines.

10. Get a bird's-eye view of the world by climbing a tree in your backyard. Dig a deep hole to discover the different layers of earth that support you and your home. Lay down on the grass and see what the world looks like from that angle. Make note of the living things that share your property.

The Ecology of Commerce: Encouraging a More Natural Economy

"... early attempts to wed green consciousness to child's play gave birth to such snoozers as juice-jug bowling pins. But just as vegetarian cuisine has evolved from penance to pleasure, the spirit of earth-friendly amusements has shifted from dutiful to delightful."
—writer Julie Bourland,
"Making Spirits Bright," *Sierra*

"Many companies today no longer accept the maxim that the business of business is business. Their new premise is simple: Corporations, because they are the dominant institution on the planet, must squarely address the social and environmental problems that afflict humankind. ... Companies are coming to realize that they may

succeed according to conventional standards and still be violating profoundly important biological and natural systems. The question is, can we create profitable, expandable companies that do not destroy, directly or indirectly, the world around them?"
—author and businessman
Paul Hawken, from *The Economy of Commerce: A Declaration of Sustainability*

Barbie will never die. *Literally.* The popular doll that has remained, for decades, a child's most precious plaything and, as well, has become a pricey collectible, is truly immortal. Mankind has finally succeeded in maintaining a species that it cannot destroy. In fact, two thousand years from now, when archaeologists dig up the landfills that have been created by mankind, what they will most likely discover, among all the other nonbiodegradables, is Barbie. Even her synthetic blond hair and rose-tinged, clear complexion will look none the worse for wear, despite her horrendous experience of such a trashy burial.

Barbie, as well as numerous action figures, water guns, and other molded toys, novelties, and sporting goods, owe the Plastics Industry *big-time* for making them immortal; in fact, nearly *seventy-one million pounds of high-density polyethylene plastic went into their manufacture in one year alone!* Plastic diapers, polystyrene cups, and other household trash also owe a debt of gratitude to the plastics industry for their indestructibility. And what better testimony is there to plastic's durability than the creation of packaging that lasts four hundred years, even though its primary purpose is to keep a product on-shelf for two months just so it can be consumed at home in two minutes!

But is such indestructibility a bad thing? Some things created by mankind need or ought to be indestructible: the aluminum siding that covers your home, for instance, the steel girders that support the high-rise building in which you work, or the rubber tires on your child's stroller. But do the aluminum cans that hold your drinks have to last for centuries after you have drained the can? Do your car's steel-belted tires have to pile up for thousands of years after they have gone even slightly bald? If you no longer have a use for it and have nowhere to store it, then what good is such indestructibility?

Have you ever stopped to think about what has happened to *all* the automobiles that have been manufactured in this country—built from steel and rubber and plastic—that have been traded in or smashed up since the early 1900s? Since the sheer number of vehicles this question represents may be difficult to fathom, consider the number of cars that you have previously owned. Where do you think they are now? Multiply this number by the number of adults in your family. Where do you think all of your family's previously owned automobiles are now? Think about where all the old motorcycles have gone, all the taxi cabs, tractor trailer trucks, antiquated farm machinery, Army tanks and trucks, jeeps, U.S. mail trucks, and milk delivery vans. Or think about aluminum cans. About how many cans of sodas, beers, juices, and seltzers do you drink in a week? A month? A year? Multiply this yearly total by the number of years you have been drinking out of cans. Where do you think all your used cans are now? Now multiply your total by the number of adults *and* soda-and-juice-drinking children in your family. Where are all of your family's used cans now? Certainly not in your home—no cabinet, closet, or cellar would provide enough storage space. You must,

instead, have transferred such quantities of household waste from within your own small environment to a much larger environment outside your home. But where, then, does this much larger environment transfer all of the accumulated mountains of waste not just from your family but also from your neighbor's family, your neighborhood's families, your town or city's families, your state's families, your country's families—from all the families of the world?

You can rest assured that all of your cans, tires, scrapped steel and aluminum, and other household wastes have been disposed of over the years—buried, heaped, plowed under, compacted, piled up, covered over, and dumped somewhere on this planet. But since your refuse is out of your sight—it is not in your back-yard and certainly not within view from your living room window—then it is most likely out of your mind. Because the natural human reaction is to avoid waste, society strives to segregate and steer clear of the waste it generates. So waste goes into an ocean, where it sinks to the ocean floor or floats somewhere else. Waste goes into a stream, where it is quickly carried away. Waste goes into the ground, where it can no longer be seen. Waste is tossed onto the side of a road while traveling to some distant destination. Waste is bundled into trash bags and picked up by noisy, smelly trucks, which then cart the trash away to some unknown locale. Waste is loaded onto barges for ocean voyages to whatever port will take the stinking cargo. Waste is abandoned deep in the woods. Waste is stashed under your theater seat. Waste is dropped on the city sidewalks. Waste is flicked out the car window or ground into the dirt. Waste is spit out, spilled out, flushed out.

Waste, mankind has learned, is not "good." It makes you wrinkle your nose, step aside, or recoil in disgust. It must, therefore, be disposed of as quickly as possible.

Even though the best way to avoid such a stinking mass of waste would be to severely restrict or even eliminate its creation, rather than go to the *source* of the waste mankind still prefers to focus on how to dispose of the even greater quantities of waste it continues to produce. And the biggest and best disposal facility known to mankind is the biosphere. After all, there is no fee for dumping and no one really complains because no one really knows where their trash goes. Even though there are ozone holes in the atmosphere, beaches must be temporarily closed when used syringes and human offal wash up on shore, birds and turtles are perishing from plastic and polystyrene ingestion, thousands of dolphins and seals are killed by viral and bacterial infections, baseball-sized tumors are found in turtles, PCBs and mercury are the newest ingredients in some of the world's finest seafoods, and crabs with nervous systems deranged by chlordane runoff foolishly try to mate with rocks on the ocean floor, the trash problem *is* being resolved. So you do not have to deal with it.

Or do you? Organochlorines from solvents, fungicides, pesticides, and refrigerants are rapidly building up in the environment and steadily accumulating not just in the food you eat and the water you drink, but, as well, in your body. Because your metabolic process has little or no effect in converting these substances into more harmless forms, they cannot be excreted. Instead, they remain in your system, stored in your fatty tissues. These compounds play havoc with your all-too-human physiology, with side effects that include cancer, infertility, immune suppression, birth defects, and stillbirths. Even though similar side effects have been observed *for several decades* in the wildlife community—in the form of decreased fertility, behavioral abnormalities, comprised immune systems, and monstrous defects that impact reproduction—it was only when similar patterns of

disruption in humans began to be noticed that the link was made between physical ailments and the slow degration of waste products in the ground and the air.

What was the response to this devastating connection made between industry and the havoc it had caused not just on the environment but also on the human species? Or, more accurately, because of a growing consumer awareness, expressed by shock and outrage, from well-documented and well-publicized findings provided by organizations such as the Environmental Defense Fund, founded in 1967, which brought to light such industry-generated environmental impacts as global warming, acid rain, and the destruction of the ozone layer? The EDF, too, sought to eliminate consumer health hazards such as childhood lead poisoning, chemicals in children's sleepwear, and asbestos in hairdryers, and industries responded quickly to this consumer "call to action." In an effort to achieve quick resolution, industry rallied to eliminate the massive quantities of waste it had generated. "Bury it!" was one rallying cry, so the waste was buried. Such sites have now been identified as toxic dumps that cry out for priority clean up while thousands of lawsuits are being argued by lawyers for unsuspecting families who live or lived in close proximity to such sites and suffer from a variety of nervous disorders, fertility problems, and cancerous growths. "Incinerate it!" was another rallying cry, so the waste was incinerated. But one study of a state-of-the-art incinerator in New Jersey revealed that burning merely altered the form of the waste: ". . . 2,250 tons of household garbage daily would annually emit 5 tons of lead, 17 tons of mercury, 580 pounds of cadmium, 2,248 tons of nitrous oxide, 853 tons of sulfur dioxide, 777 tons of hydrogen chloride, 87 tons of sulfuric acid, 18 tons of fluorides, and 98 tons of particulate matter small enough to lodge permanently in the lungs. . . . For every 100 tons of trash, incinerators produce 30 tons of fly ash, a granular substance that contains most of the toxins from paint and plastic, as well as mercury, lead, cadmium, and other heavy metals. The fly ash is then trucked to a landfill where it has to be enclosed in plastic liners for many thousands of years. The plastic presently being used in fly-ash landfills is guaranteed for only twenty years, and landfills containing toxic fly ash in New York and New Jersey have reported leaks within months after installation."

If only Barbie could be called in to plug the leaks. Or, better yet, if only Barbie could be fashioned into a trash-eating version of Pac-Man so she could gobble up all the toxic wastes, household refuse, and heavy metals and then hold them safely for an eternity within the confines of her curvaceous, polyethylene plastic, indestructible body.

The task of eliminating what mankind has already eliminated has gone well beyond the level of sheer frustration, well beyond the scope of desperation, and has now pushed the life-or-death envelope, hitting panic buttons everywhere. A citizen of Allen County (Ohio) who helped to organize Citizens for the Environment to protest the chemicals that were being released into the water and atmosphere by a nearby British Petroleum refinery reports that he has health problems and, as well, his young daughter has serious respiratory problems. Then he sighs and slowly opens his hands, one finger at a time, as he reports how others who live nearby have fared. "The guy across the street has cancer," he begins. "The woman down the street has brain cancer. The woman around the corner has brain cancer. The woman who lives next door to my child's friend has cancer. The woman on the next block has breast cancer. The guy next door to her has cancer. And so does the woman next door to him.

Those are just the homes I can see when I am looking out my own front door." While British Petroleum's future emissions into the environment can be regulated— and, perhaps, over time, even eliminated—while the air can eventually become clearer through reduced atmospheric emissions, and while the water can eventually become purer with less particulate matter falling from the sky, the cancers, unfortunately, are still operating under "the green light." They will continue to spread and grow.

While it may be all well and good to recycle your aluminum cans at work, to plant a tree in your city park, to bake in reusable metal or glass pans rather than aluminum, or to feed your dog or cat dry food rather than canned, such efforts, while well-meaning and oftentimes personally reward-ing, are, in terms of eliminating waste in the environment, like try-ing to extinguish the infamous Chicago Fire with a teaspoon of water. And even though millions of other people may be contributing their own teaspoons of water to quell the inferno, the fire is still raging way out of control. "In Hanford, Washington," writes Paul Hawken in his book *The Ecology of Commerce,* "the site of a nuclear weapons research facility managed for the U.S. Government by General Electric, there are enough wastes stored there to cover all of Manhattan with a radioactive lake forty feet deep. Some of these wastes are stored in underground tanks. The tanks are leaking."

For years, industry's only solution to reducing the massive quantities of waste that has accumulated over the years has been to simply clean up the waste. Since 1970, the United States has spent over *one trillion dollars* to monitor, litigate, contain, and curb pollution and hazardous waste. Of an estimated ninety thousand hazardous waste sites, twelve hundred have been designated as priority clean-up areas under the Superfund law.

But as precious time is being spent cleaning up priority areas in a long, slow, tedious process that requires careful handling and re-disposal, other sites reach the end of their "safe-storage time" and begin to release their own organochlorines into the atmosphere and ground water. When one barrel starts to leak, it is not just the immediate environment or nearby living things that are placed in jeopardy; the air and water play a tremendous role in making the effects even more far-reaching. For instance, during a 1992 international conference on herpetology (the study of amphibians and reptiles), participants were alarmed to discover, after they informally pooled their findings, that their evidence was irrefutable: frogs were disappearing from the planet at an alarming rate. But not only were frog populations being decimated in known industrial waste regions, but also in pristine wilderness locales where there was ample food and no known pollution. If such endocrine and immune system failure was occurring in frogs—in the lower levels of the animal kingdom—the conference participants wondered, how safe was mankind from experiencing a similar fate? And if not just the waste site regions but *all* regions of the planet were being affected by mankind's toxic accumulation of waste, then was mankind slowly strangling itself and its own planet?

While it is certainly laudable that large-scale, planetary clean-up has become a priority—after all, waste not only affects every neighborhood but also every ocean, every lake, every forest, every field, and every continent—and while environmental clean-up has contributed in positive ways to the economy through the growth industries it has created, what really needs to happen is a two-pronged attack upon waste: to eliminate the waste as well as eliminate *the creation* of it. This means that businesses must not just add value to the economy and society, but, as well, add value to the environment by protecting and restoring it to its original form— pure, clean, and pristine.

Industrial ecology ought to be included in the charter of each and every yearly business plan, whether the business is large or small, old or new, product-oriented or service-oriented. Then, and only then, will businesses begin to accept both responsibility for the degradation of the environment and for its restoration. In accepting such responsibilities, the ensuing actions they take will ultimately help to change the world and the way each individual lives in environmentally healing and enhancing ways.

Consumable goods, by definition, are those products that are designed to be used and consumed, usually only once, and then, when discarded as waste, are considered to be *wholly biodegradable*. A biodegradable product, by definition, is capable of transforming itself into "food" for another organism with no toxic residue that would cause harm or have any accumulative effects during its breakdown. In essence, the environmentally "perfect" product is one that is capable of turning itself back into life-sustaining dirt without harming either the environment or any other living thing in the process of its decomposition.

The best example of such a consumable good would be a ripe, unsprayed apple. You, as the consumer, purchase the untreated apple. You eat everything but the core and the stem. You discard this "waste." A passing skunk or raccoon may eat your apple core, or your apple waste may eventually soften and break down in the ground. Like the forest which, if left untouched, constantly thins and "feeds" itself with its own consumable products and waste, mankind needs to learn how to generate its own consumable products not just to "feed" itself but also to "feed" the environment. Food that has been tainted with pesticides is not consumable or biodegradable. Clothing that has been made from natural products, such as cotton and silk, and then treated with zinc and tin to give it a heavy, luxurious feel is not consumable or biodegradable. Your television, which contains some four thousand chemicals, ten to twenty grams of mercury, and an explosive vacuum tube is not consumable or biodegradable. Walk through any shopping mall and look around you at the mountains of nonbiodegradable, nonconsumable goods that are being sold and which, in a miracle that can only be performed by mankind, are constantly replenished in a seemingly endless quantity. What you are seeing, in essence, is next week's environmentally harmful trash pile—all packaged up and ready to go. What will *you* choose to purchase that will eventually contribute to that already enormous trash pile?

During the Depression, a Farm Chemurgic Council, which included such revered businessmen and ingenious inventors as Henry Ford and George Washington Carver, strived to apply farm products to industrial uses in an effort to create truly consumable consumer goods. In 1941, Ford's efforts paid off when he designed a prototype automobile that boasted a body made of soybean plastic, an engine powered by ethanol (a corn-based fuel), and tires made from goldenrod. In effect, Ford literally "grew" his own car—a vehicle which, when discarded, would decompose into dirt in his own backyard and serve as organic fertilizer for his next homegrown car. Ford was so confident that oil prices would rise after World War II that he envisioned a world in which all automobiles would have to be so "grown." Although the first part of his prediction proved accurate, auto manufacturers kept the farmers out of the auto-making business by convincing the government and American consumers that their mobility as well as the nation's economy depended upon oil, steel, plastic, and rubber—without

regard to the impact such products would have on the environment.

"Two of Huffy's bicycles give new meaning to the word *recycle,*" writes Julie Bourland in her article about the efforts being made by some toy manufacturers today—like modern-day Henry Fords—to reduce waste in their products. "The frame of the Metaloid is made from one hundred and twenty aluminum cans and industrial aluminum scrap. Huffy's Eco-Terra salvages material in an even bigger way, using plastic containers and a percentage of reused steel for the bike's pedals, grips, handlebar, and frame. The manufacturer of the archetypal little red wagon, Radio Flyer, is also hitting the environmental trail with the Earth Wagon, which boasts wood-like planks cunningly formed from old plastic jugs." The Nature Company sells a Frisbee model molded from sixty percent recycled polyethylene plastic. And Holbrook-Patterson Inc.'s Noch-Blox building logs are made of plastic drink containers; a pound of their blocks is equal to ten plastic milk jugs that would have been preserved for all eternity next to Barbie in the local dump.

But children's toy companies are not the only businesses that are making grand efforts not just to limit or reduce wasteful products but also to eliminate them. Lawn care companies, for example, once told their customers, "A weed on the lawn is like a coffee spill on your living room carpet. If you use our [now outlawed] herbicide your lawn can be a perfect green carpet." The "lawn talk" these companies now give their customers is more about sustainable gardening, less fertilizer, and little if any pesticide. The lawn has evolved from a social statement—"See, we take care of our weeds and pests"—and a matter of civic pride—"My lawn reflects the care homeowners put into their properties in this town"—to free-growing herb and wildflower plots and mowed greenswards, also known as "free-dom lawns," which oftentimes include different varieties of grass as well as ivy, ajuga, moss, violets, clovers, and even the once greatly despised dandelion, all of which offer a uniform carpet of green when mowed but then turn into an esoteric mix of texture and colors when resprouting.

As well, some communities that have town pools have entered into the business of environmental awareness by experimenting with chlorine reduction not just because of the discomfort it causes swimmers but also because many organizations, including Greenpeace, contend that chlorine from pools, industry, and other locations is contributing to depletion of the atmosphere-protecting ozone layer. (In fact, chlorine is an element that is extremely unstable and volatile, ready to recombine with other elements into molecular compounds that are almost universally poisonous to invertebrates, plants, animals, and humans.) Brookline, Massachusetts, for example, has sharply reduced their use of chlorine in the community's Olympic-sized swimming pool, diving pool, and children's pool in favor of using an ancient bacteria-killing practice known as ionization. Today's ionizer technology, which uses an electric device that dispenses ionized copper and silver into the water to kill bacteria while keeping chlorine levels at a low 0.4 parts per million (representing an average seventy percent reduction in chlorine), dates back six thousand years to ancient Greece, when royalty discovered that water tasted better and seemed more healthful when dispensed in silver pitchers and goblets. Centuries later, during the Western expansion of the United States, Conestoga wagon operators dropped copper pennies into their water casks to keep the water potable during their long journeys. And, more recently, NASA perfected the ionization process during the Apollo moon launches to purify the astronauts' water.

While Brookline's recreation director estimates that little more than a two thousand dollar savings a year will be directly realized through the ionization program, he is ecstatic not just about the reduction of the use of the environmentally controversial chemical, but also about the drastic reduction in nonbiodegradable wastes that Brookline once contributed to landfills. Because chlorine is highly corrosive, especially to pool structures and machinery, the pool and its equipment can now be preserved for a much longer period of time; the handrails and lifeguard chairs will not have to be replaced so frequently. "If you can add two or three years to the life of your machinery and the structure, you're saving thousands and thousands of dollars," reports Thomas J. White, president of Crystal Water Systems, which offered the twelve-thousand-dollar Crystal Water unit to Brookline for nothing and will cut the

Central Park, NY, 1956 © David Vestal 1997

purchase price in half if Brookline officials like the end results.

Today environmentally savvy companies, most of them small-scale operations, are springing up from some of the most offbeat ideas and are challenging larger-scale companies to follow suit with similar environmentally safe products and manufacturing procedures. As well, they are also changing the way consumers think about the products they are purchasing and the people (not just the companies) from whom they are purchasing product. Take, for example, the group of women who started Mountain Women Soap. Each of the four women was labeled a "slow learner" in school. Each had dropped out of school before the tenth grade. Each had become single mothers whose relationships with men had often been abusive. Two had committed petty crimes. All had eventually gone on welfare, collecting state aid for almost two decades.

The four women met while taking a course for high school equivalency diplomas. They were sick of being broke, sick of wondering where the next check was coming from, and, most of all, sick of being so dependent upon state aid. One day, one of the women came into class with a bar of soap she had made from an old family recipe. Although she was considered "artsy," her soap turned out to be more than just another neat, crafty thing to be admired. Three other women saw the possibility of a business in her bar of soap—a way, so to speak, to make a "clean break."

In honor of their rural roots and the White Mountains where they live, the four women named their company Mountain Women Soap and set about producing all-natural, animal-free soaps (without lard, a product in most commercial soaps that does not break down in water) in assorted shapes, scents, and sizes. The company has been housed in a variety of locations over the years since its inception—from a church anteroom to the back of a karate school. But in 1996, the company grossed nearly fifty thousand dollars. "A hand-lettered scrap of paper taped to the door of an old cabinet maker's shop marks its current headquarters," writes Judith Gaines in her article about the environmentally clean, soap-making company. "Inside, the women make the soap in large baking pans, shape it in handmade wooden molds that have to sit for three days, and then package it themselves. Hearts, balls, and bars of soap sit atop tables, racks, counters, and bookshelves, as coconut, jasmine, vanilla, lavender, and other fragrances waft from room to room."

Another unique, environmentally savvy business, Conversion Products Inc., attracts a great deal of attention in trade shows when it displays its most popular product, the Adirondack chair, with this accompanying sign: *Sit on Me! I'm Made from 240 Milk Jugs!* "The recycling part really caught people's attention,"

says David Sparks, owner of the company that has built its business around products created from "plastic" wood. Although the traditional customers of Conversion Products Inc. have been such "heavy-users" as cities and towns, schools, golf courses, zoos, and housing authorities for the company's durable plastic wood park benches, picnic tables, and trash receptacles, the company's recent foray into home and garden plastic lumber products has attracted a large number of consumers who are not only impressed by the classic design, rugged construction, and comfort of the Adirondack chair, garden bench, planters, and flower boxes, but also by the low maintenance. Because plastic wood is impervious to water, salt, oil, and chemicals, it will not rot, splinter, crack, fade, or need to be painted year after year.

"I'd always wondered where all my recycling was going," says Conversion Products customer James Linderman, who lives on Sip Pond under the shadow of Mount

Monadnock in New Hampshire. "So I was delighted to see my milk cartons reemerge as furniture. I bought three Adirondack chairs and stools and put them out by the pond. They're very comfortable, well designed and heavy, which is a plus here where we get tremendous winds. I also don't have to worry about painting them, I can leave them out in the sun and I'll probably leave them out all winter."

In combining an environmentally sensitive consciousness with creative furniture design, Conversion Products Inc. is just one of many small businesses that is providing a positive power of example to other companies and skeptical consumers that recycled products are not always synonymous with junk and that good products can be created in innovative ways that also reduce the world's waste.

The conventional wisdom is that humans are bad for the environment. But if mankind just sits back and accepts this without argument, as if it were cast in stone, then there is no hope for humanity. The planet is surely doomed.

But what if such wisdom could be proven wrong? What if humans could actually be *good* for the environment? What if industries were able to replicate more of the "natural world" in their manufacturing—a world in which nothing goes to waste, in which each thing that is produced is used or is vital to something else—whether that something else is a living system or simply a collection of nuts and bolts?

Industrial ecology is proving itself to be more than just a "what if," more than just a fantasy, more than just a possibility. More and more, the working world is combining commerce with conservation, profit with preservation, financial security with planetary sustainability. The environment, as well as business, are together forging a valuable and vital restorative relationship with one another that ensures and safeguards the continued survival of each.

Some Ways to Encourage a More Natural Economy

1. It is not always necessary to fertilize a lawn to keep your lawn green. Natural organics and sulfur-coated urea provide a more uniform, longer release of nitrogen than faster dissolving inorganic chemicals, which leach out into the ground water. As well, there is evidence that returning grass clippings to the lawn equals the nutrients of two fertilizings. Use a mulching lawnmower or run over grass clippings again and again with a non-mulching mower until the clippings are small and almost disappear.

2. Read labels to find out what ingredients are being used in the products you purchase. If you are not familiar with any of the ingredients, learn what they are. Sharply curtail purchasing products that contain nonbiodegradable elements. Make sure that those things that are being presented as all-natural, preservative-free, or organic really are.

3. Remember that whatever you spray into the atmosphere—even if the "atmosphere" is in your own home—must go somewhere. Just because you no longer see it does not mean it has disappeared harmlessly. Examine your use of hair sprays, household air fresheners, colognes, perfumes, shoe and boot leather preservatives, paint thinners, and so on. Consider non-polluting alternatives. For example, a scented candle or a mixture of herbs boiled in water can freshen stale household air just as well as a household air freshener can.

4. Whenever possible, purchase products that have been hand-crafted or handmade. Although such items can cost more than those that have been manufactured—a handmade canvas-and-wood canoe is far more expensive than a manufactured fiberglass model—the handcrafted canoe has contributed very little to the destruction of the planet.

5. Buy in bulk whenever possible. Stock and refill canisters with

tea, flour, sugar, coffee, spices, pastas, and other items purchased from the bulk sections of stores rather than contribute empty containers to landfill piles.

6. Find out how local businesses are recycling within their own manufacturing process and the ways in which they are reducing or eliminating the creation of waste product within their own business.

7. Join an organization such as Business for Social Responsibility, the Social Venture Network, the Environmental Defense Fund, the Council on Economic Priorities, the Coalition for Environmentally Responsible Economics, or a local group of corporate ethics consultants.

8. Find out where your previously owned automobiles have gone as well as your aluminum cans, your broken down VCRs, your old record players, used batteries, 8-track tapes, old video game machines, the transistor radios you have owned, and so on.

9. Support manufacturers that offer "buy-back" or "trade-in" programs that encourage you to return an appliance to the original manufacturer when you no longer have a use for it or desire a new one.

10. Live and abide by carry-in/carry-out ethics in the outdoors; encourage friends and family members to do the same. The advice to "leave only footprints" sharply reduces litter and refuse, which is becoming increasingly more prevalent on popular beaches, hiking trails, and mountainsides. In fact, in 1996, sherpas hauled out *forty-four hundred pounds of trash* from the South Col route on Mount Everest. The trash, some of which dated back to 1952, included empty oxygen cylinders, cook stoves, tents, ropes, ladders, batteries, and clothing. Despite this incredible effort, an estimated thirty-three thousand pounds of discarded gear—and ten frozen bodies—still remain. Wherever you go, diminish existing trash by carrying out things others have carried in. While that may not be your "job," it is vital to *your* planet.

You Call This School?
Educating with Nature

"Hoping to help our daughters develop an appreciation of nature, my husband and I took them camping. After we arrived at the campsite and unloaded our gear, we all set to work. My husband had the girls gather pine needles to make soft 'beds.' Then he instructed them to bring rocks to form a circle in a cool stream. We placed our perishables inside. My husband explained that this would be our 'refrigerator.' Next, we made a 'stove' by making a ring of stones topped with a grate. The girls were impressed and excited. Our five-year-old smiled up at her father. 'If we get more rocks, Daddy,' she asked, 'will you make a TV?'"
—contributed by Barbara Bock to *Reader's Digest*, June 1996

When a hundred and one ninth-grade boys and girls tumbled off school buses on a rainy September day and peered out into the dripping, dreary-looking, fog-shrouded, wooded landscape near New Hampshire's Mount Monadnock—a place they would be calling "home" for the next eleven nights—they had to wonder what had ever possessed their parents to place their lives in such jeopardy. After all, they were students at the largest independent day school in Massachusetts and, individually, worth about $17,600 in tuition fees alone. Most lived in affluent suburbs outside Boston, where the most challenging outdoor experience was negotiating barely plowed sidewalks in the winter; the wildest creature with whom they had ever made contact was the gray squirrel. Make-up, nice clothes, CD players and stacks of CDs, a comfortable bed in their own spacious bedrooms, an endless supply of pizzas and sodas, cellular phones, video games, e-mail accounts, shopping malls, and country club memberships were more in line with what their lives were really like than what they were about to face: an under-the-canvas, no-matter-what-the-weather wilderness bivouac that was *required* for graduation from the prestigious Buckingham, Browne & Nichols School of Cambridge, Massachusetts.

It was that simple: no bivouac, no diploma. What had started in 1951, when the school was all-male and known as Browne & Nichols, as a sort of "Davy Crockett" journey into the wilderness for the city lads, a time in which they made ax handles, caught fish, and stocked their "wild foods" kitchen, had evolved over the years into a '90s-style "Biv" that still valued self-sufficiency and resourcefulness—participants had to pitch two-person tents; build A-frames from narrow tree limbs, tarp, and ropes; dig latrines; bathe in frigid Silver Lake; cook and eat their own food; climb Mount Monadnock; shimmy up ropes; tie ropes with a variety of knots; and scale rock walls—but also offered courses on diversity and human relations and emphasized cooperation and teamwork.

"I'm in denial right now," declared Dimitri Zagaroff of Cambridge at the start of his 1996 Biv. "I simply don't believe I'm here for the next two weeks. And I have no idea why I'm here, except to be in the woods. And to bond, I guess." Dimitri, shy and often reluctant to make friends, had squirreled away two books by German philosopher Frederich Nietzsche in his gear, hoping to bury his head in a book and keep to himself on his requisite outdoor adventure. But, surprisingly, by the afternoon of the first day, as Dimitri and his squad of six other students were struggling to lash together an A-frame, grunting and groaning and giggling as a group, he had worked up the courage to turn to a boy who was standing next to him and say, "Oh, your name's Brian? Pleased to meet you, Brian." Even more surprising, several days and nights later, on the eve of the bivouac's end, Dimitri sat alone in his tent and shook his head. "I'm in denial about going home!" he grinned. During the Biv, Dimitri had turned fifteen. "It rained and my tent leaked," he recalled. "I didn't get any presents and I couldn't call my parents. But I made it through. It was a great feeling, and it made it the best birthday of my life." Through his experience, Dimitri realized that he had learned just as much about human nature as he had about Mother Nature. "For the first three days, we just put everything off and figured someone was going to do it for us," he explained. "Then most of us realized it wasn't going to happen

unless we did it ourselves. We learned to rely on ourselves."

Other startling epiphanies of self-discovery occurred in both the boys and the girls during and after the Biv. Alexandra Dingman, for example, had arrived at the Biv declaring that she would rather spend her nights in a motel; months later, she is still quick to remind people that she *endured* the Biv and, of course, *did not* have fun. But she, like Dimitri, learned a great deal about herself and others."At school you see certain girls with lots of makeup, with their hair done up, with expensive clothes," she explained. "Here you see them in their grandfather's work shirt and boots that don't fit. And when you live with them, you find out they're just like you except for the clothes and the makeup. Everyone here looks the same, so you make judgments on personality rather than looks." Ashley Dorian, who had brought along enough clothes to outfit her entire squad and thought "we'd have tents that just kind of blew up" rather than the two-man metal pole and canvas tent that she and her tentmate had to struggle for hours to put up, recognized by the end of the bivouac that she was clearly living "a pampered life." As she vigorously scrubbed a plastic bowl at her squad's campsite, she talked of returning to future Bivs as a student staffer. "This has been good for me," she remarked. "I don't do many chores at home. I never, *ever,* do dishes. But this morning I spent three hours—three hours!—scouring a pot with S.O.S. stuck on sticks. I'm proud of that, and I'm proud of what I've done."

Other students learned a great deal about what they could do in the outdoors. Erica Birmingham, daughter of Massachusetts Senate President Thomas Birmingham, stood in awe in front of the A-frame shelter that she and her seven- squad team had just roped together. "I never thought this group was going to be able to build something like that out of tarps, and wood, and rope," she exclaimed, her eyes wide. "I mean, we didn't even use any *glue.*" And Elizabeth Gates, who had initially fretted before the Biv, with a tinge of desperate anxiety in her voice, that she was certain "they're going to make me climb some sort of tree," later raved about the required Mount Monadnock climb. "The view was unbelievable, like nothing I've ever seen before," she gushed. "I think I learned a lot about myself here. And one of the things I learned is that I'm not as much of a city kid as I thought. I learned I can handle the wilderness."

By the end of Buckingham, Browne & Nichols' 1996 Biv, fifty-four of the kids had "soloed," or chosen to spend one night by themselves a mile or so from the group. Fifty-three had earned the title of "100 percenter," meaning that they had each bathed in the chilling waters of Silver Lake every day. Not one student had needed to visit the local hospital. And everyone had endured nearly ten straight days of rain, two inches of water both inside and outside their tents, and sopped sleeping bags.

As well, many of the students had come away from the experience realizing—perhaps for the first time—that there were many simple pleasures in their daily lives that they often took for granted: hot showers, flush toilets, warm clothing, parent-cooked food, listening to music, sleeping in a dry bed. They learned that survival was more than just having to deal with not having the material possessions to which they had grown accustomed, but making use of what they did have. They learned that sometimes interactions with others were difficult but that later on in life, when they would have to get along with people they might not like, they could rely upon the same cooperative spirit that had helped their squad to work together. They learned how to tie a simple square knot, build a crude shelter, and cook outdoors. They learned what the view is like from the top of a mountain and how weather really can dictate mood.

One exhausted Biv summed up her feelings about her bonding experience with nature, self, and others as she boarded the bus that would take her back to her city home. "You spend a lot of time being sad here, and a lot of time being happy," she said with a sigh. "When it's raining day after day and you're soaked and miserable, you think it's never going to get better. Then the sun comes out and you feel really great. Your spirits lift and you forget what was so bad about the day before. And even though you miss some things that are in your own world, at home, you discover that there are lots of other things out here in the outdoors that are really kind of cool. I think I'm actually going to miss some of those things when I return home."

And yet how many of the Biv participants do you think would have *elected* to go out into the wilderness had they been given the choice?

How many of the Biv participants do you think would have been encouraged by their parents to undertake such a difficult experience? Education today stresses how to prepare for life outside of school not in the great outdoors and not in a cooperative venture but in the stressful, competitive, all-work-no-play world that focuses on individual effort, emphasizes fat pay checks and impressive job titles, requires playing the corporate-politics game, and necessitates considerable knowledge in such fields as economics, computers, telecommunications, and the sciences. So the classroom emphasis in educating students, even in the early grades, is on such subjects as math and computer science and on the development of individual skills and talents. As a result, schools cry out for computers and textbooks, not for funding for field trips to the mountains or the wilderness. Schoolteachers have their hands full trying to control overcrowded classrooms comprised of students who have a diversity of learning levels and learning interests; they have little time to encourage study of even the simplest ecosystems that thrive right on school property. School administrators are under constant pressure from parents who want guarantees that their children will graduate high school with more than just a third-grade reading level and will, as well, be capable of stringing together a paragraph that is not just grammatically correct but also makes sense; thus, they have little time and even less inclination to entertain the possibility of adding any school programs that deviate even the slightest from those that provide measurable, testable outcomes. Politicians who eloquently promise to protect and promote the education of America's children are often more concerned with how American children compare with the rest of the world's children not just in reading, writing, and arithmetic, but also in calculus, biology, engineering, astrophysics, economics, and trigonometry; thus, they talk little about how the future sustainability of the planet and all of its living things will depend upon the interest and motivation today's children have in saving it.

In sum, the value of education, in the minds of those who are most responsible in bestowing this education upon children, lies predominantly in the presentation of and constant drilling in facts and figures, not in the development of observations and interpretations. "Children enter school as question marks and leave as periods," writes author and educator Neil Postman about the practice of diminishing the curiosity-wonder that is innate in children. Children, who are born as lively little questioning machines who marvel at the world in which they live, are, through the education process, transformed from very young ages into robotic responders whose sense of wonder is replaced by practical and applicable knowledge. Even with only a minimal amount of schooling, it is not long before some of the most open-ended, fascinating questions of youth—"Is the sea alive?" "What songs do the birds sing?" or "Why does chocolate taste happy?"—are replaced by questions that demand just one right, inarguable answer—"What is the capital of South Dakota?" "What is the temperature at which water freezes?" or "How many acres in a square mile?"

Pablo Picasso once pointed out, "Computers are useless. They only give you answers." What is more significant, then— an education that stresses seeing what everybody else has seen, learning what everyone else has learned, and thinking what everyone else has thought, or an education that allows students to see what everyone else has seen but, as well, encourages them to think what nobody else has thought before about such things by allowing appreciation for such things, by instilling awe and curiosity about them, and by celebrating the spontaneous discovery and creativity that can result from the experience of such insights?

American poet and writer Wendell Berry once made this observation about education: "And we will know that of all the issues in education, the issue of relevance is the phoniest. If life were as predictable and small as the talkers of politics would have it, then relevance would be a consideration. But life is large and surprising and mysterious, and we don't know what we need to know. When I was a student I refused certain subjects because I thought they were irrelevant to the duties of a writer, and I have had to take them up, clumsily and late, to understand my duties as a man. What we need in education is not relevance but abundance, variety, adventurousness, thoroughness. A student should suppose that he needs to learn everything he can, and he should suppose he will need to know much more than he can learn."

How, then, can today's teachers combine knowledge with wonder in creating more of an experiential education in which some truths are discovered not just because they are true or understood to be true but because they are *experienced* as being true? A teacher, for example, can teach his or her students the simple fact that maple trees have leaves that change color in the fall. But, as well, that same teacher can also teach his or her students how to experience—not just learn—this simple fact. One way a teacher can do this is by having the class observe for themselves the maple trees in their own schoolyard and to become "field observers" by taking notes on when they first notice color transformations. Another way is to take the class on a trip to a local arboretum, where they may not only be able to confirm what they have already learned in the classroom to be true—that maple leaves change colors in the fall— but, as well, can expand upon this knowledge base by learning and observing that *each species* of the maple tree changes leaf color at a *different* time of year (the Norway maple tree turns color very late in the fall, for example, while sugar and red maples turn in early October). The astute teacher can go even further and present a "what-if" scenario to students that encourages valuable critical-creative thinking—a way of thinking that is not only vital in problem solving and conflict resolution but also in innovative and creative conceptualization—by asking students to compare city maple trees to country maple trees. "Will it make a difference," the teacher might ask students, "in the time at which the leaves change color or their brilliance or how long the leaves stay on the tree before they fall to the ground if the city tree is more stressed than the country tree because its roots might be bound-up and water might not be as readily available?" This experiential type of learning—absorbing

facts as well as applying them through the experience of first-hand observation—expands the classroom learning experience well beyond the boundaries of four walls and creates a truly sensory (and much more impressionable) *involvement* with learning. This fosters the develop of a *relationship between the knowledge and the student and creates an intimacy with what has been learned;* as a result, the student *wants* to take this learning out of the classroom and into his or her life outside school by observing trees in the backyard or by taking an exploratory nature walk around the neighborhood.

The deeper the impression that can be made through such a unique and personal learning experience, the more long-lasting such learning can be, sometimes staying with students well beyond their classroom years or even long into adulthood. Because such a teaching approach imparts facts as well as encourages children to "think beyond," it empowers children to take charge of their own learning. And this combination of information and empowerment is truly what teaching should be all about, for, in the end, what results is the creation of *lifelong learners*.

How can teachers transform ordinary school days and lesson plans into an extraordinary lifelong learning experience for their students—an experience that encourages them not to always rely upon what they read and what they hear, but on what they can experience for themselves so they can apply this experience to a better understanding of themselves and the human and natural worlds in which they live? While several colleges across the United States offer their freshmen experiences in the wild, some secondary schools offer experiential trips, and Outward Bound adventure programs are geared to junior high and high school students, children of *all ages* ought to be given opportunities for hands-on experiences both inside and outside the classroom *as soon*

as possible. This does not mean that children should have to wait until their summer vacation in order to participate in out-of-the-classroom learning experiences provided by a summer course, a family camping experience, or a summer camp program that may house the children in cabins in the woods but focuses instead on instructional learning—astronomy or computers, for instance—or attaining goals—such as losing weight or improving a tennis serve. Ordinary school days need to be transformed into challenging, lively learning sessions that encourage children to ask, "Why should this be?" and then give them the hands-on experiences that will help them to discover a variety of possible answers to such a question.

One of the most profound examples of such a life-enhancing educational experience occurred nearly forty years ago for a young boy named Geerat Vermeij, who is today a renowned evolutionary biologist and paleontologist at the University of California at Davis. When Geerat graduated from Princeton University and set his sights on pursuing doctoral studies in biology and geology at Yale University, he knew the odds of his being accepted into the program were slim. Geerat had been blind since the age of four, and he knew that the person who would be interviewing him would doubt Geerat could stay on top of the vast scientific literature that was being produced because nothing was being published in Braille. Too, he knew his sightlessness would be perceived as a detriment in his work in morphology, or the study of the form and structure of organisms, for it required close observation of minute details.

Geerat sweated through the interview, hearing the skepticism in

Fire Island, NY, 1956 © David Vestal 1997

his interviewer's voice right from the start. But Geerat enthusiastically recounted his experiences at Princeton and, as well, his long-time dedication to the study of shells. At the conclusion of the interview, Geerat could tell that Yale, his first-choice for study, would most likely not be in his future. But he tried to keep his spirits up and his voice full of optimism. When the interviewer asked if Geerat would like to "see" Yale's mollusk collection, Geerat eagerly accepted.

"Here's something," the interviewer said as he handed Geerat a specimen. "Do you know what this is?"

Geerat's fingers raced around the shell. His delicate fingers registered the widely separated ribs parallel to the outer lip, large arperture, low spire, and glossy ribs. "It's a *Harpa*," he answered. "It must be a *Harpa major.*"

"Well, now, that's correct," replied the interviewer, his voice tinged with surprise. "Now how about this one?"

Geerat took the new shell in his hand. His fingers traced the smooth, sleek, channeled suture, the narrow opening. "I'm pretty sure it's *Oliva sayana,* the common one from Florida," he replied.

For a brief moment, the interviewer said nothing. Then, with his voice full of enthusiasm, he promised Geerat his support in coming to Yale; Geerat was later accepted into Yale University with a fellowship. But Geerat owed a debt of gratitude not to his professors at Princeton or the interviewer at Yale whom he had impressed but to his fourth-grade teacher, Mrs. Colberg, and some shells she had brought into the classroom one day from a trip she had taken to the west coast of Florida.

Because the young Geerat had to rely upon sounds, odors, shapes, and textures to learn about the world around him, Mrs. Colberg's elegant Florida shells were like nothing the ten-year-old boy had ever felt; they triggered a surge of curiosity that drew him to examine the shells whenever he could. "Mrs. Colberg's finds felt as if they had been crafted by a sculptor with an eye for intricate detail," Geerat recalled in his book *Privileged Hands: A Scientific Life*. "How could one explain a shell as odd as the lightning whelk, with a spiral crown of knobs at one end and a drawn-out spout at the other? Why was its interior so stunningly sculpted with smooth, evenly spaced ribs that spiraled away beyond the reach of my fingertips?" Mrs. Colberg told Geerat of the magnificent beaches where she had gathered such "works of art." Classmates brought into class shells they had gathered from their family trips for Geerat's inspection and told him of the warm-water beaches they had visited that had gently polished and then deposited the exquisite shells on sandy shores. "My fourth-grade teacher had not only given my hands an unforgettable aesthetic treat," Geerat wrote, "but she aroused in me a lasting curiosity about things unknown. Capturing

the essence of her task, she created a freedom for me to observe, an encouragement to wonder and an environment in which to ask a genuine scientific question. Once aroused, my curiosity knew no bounds. I wanted shells of my own, and I longed to know the name and habits of the animals that built them. Mrs. Colberg gave me a few of her prize specimens, and when another class visited the American Museum of Natural History in New York City, they returned with a box for me, full of the most marvelous shells." Geerat's brother read scientific books out loud so Geerat could transcribe, in Braille, information about each shell for future study and reference; his father built a sturdy cabinet where he could store his treasured shells and books.

Both inside and outside the classroom, Geerat received unanimous encouragement. Not one person ever told Geerat that he could not pursue a career that would turn his curiosity about the world of organisms into a lifelong adventure of learning that would take him back into the classroom as a teacher himself and, as well, out into the field, to the beaches that Mrs. Colberg had once encouraged him to dream about from his seat near the window (and near the shell collection) in the East Dover (New Jersey) Elementary School.

There are many innovative, hands-on, experiential educational programs geared to children of all ages that are springing up across the country. Some start from within the classroom, sparked by enthusiastic teachers who recognize the value of creative lesson plans that encourage students to look beyond the textbook, the calculator, and the computer. Some start from excited students themselves, whose insatiable curiosity has been both praised and promoted by their parents, teachers, and guardians. Some start from nurturing parents, who not only encourage those who teach their children to be creative and innovative but oftentimes provide these educators with some of the tools needed to expand the mundane classroom teach-and-test experience—through subscriptions to magazines, sponsorship of field trips, show-and-tell presentations to the class, or supplying materials that will complement classroom study. And some start from organizations outside the school systems that seek to expand the horizons of students in order to provide them with valuable life lessons while, at the same time, the organizations benefit from receiving valuable assistance that helps them to achieve their goals.

One such program links lower-income urban youngsters with their suburban and rural counterparts in the common purpose of planting and harvesting fresh produce for homeless shelters and soup kitchens. The Food Project, a nonprofit organization based in the wealthy Boston suburb of Lincoln, emerged after a question was posed that was based on both wonder and curiosity. "What if," the question began, "Boston's youth were brought together to grow food for the hungry? Would something profound and enduring result?" Five years later, the Food Project is flourishing, spreading the children's perennially successful, organically grown harvest to such grateful organizations as the Boston Food Bank, the New England Shelter for Homeless Veterans, Food for Free, and Rosie's Place, a shelter for women and their children. Led by older volunteers, the children who participate in the Food Project learn more than just how to grow and harvest their vegetables. ("When you put the potatoes in the crate, try to be really careful," cautions one member of the farm picking group to new

near Tuxedo, NY, 1956 © David Vestal 1997

volunteers. "Those potatoes are like babies; they need care.") Participants also learn how to work through their racial, social, and economic differences—skills that are necessary to life outside the classroom but which might not have been as effectively taught within the confines of the classroom. Too, the children learn how to pull their own weight within a cooperative group venture, how to respond to others who slack off by channeling their anger in effective ways, and how their combined efforts, in the end, affect the world outside their communities.

"I like it here because I get to pick stuff and take it back to people who need it," explains Ceneio, a fifth-grader from the inner-city locale of Dorchester. Her only

Itapoã, Bahia, Brazil, 1961 © David Vestal 1997

prior exposure to dirt had been when she would help her aunt plant flowers in a pot. As Ceneio rests on the ground on her knees, she bends over and runs her glove-less, dirt-caked hands through the rich soil, searching for any potatoes she might have missed. Then she glances over at the crate of potatoes she has just filled, and her face radiates pride. "I think there's enough potatoes for a thousand people!" she beams. And yet, more important than feeding "a thousand people," Ceneio and hundreds of other children volunteers have learned not just how to plant seeds, pick carrots, or carry crates loaded with produce, but, as well, have learned how to make connections with nature, themselves, and one another—a

Willow Creek, CA, 1959 © David Vestal 1997

learning experience they will use throughout their lives.

Henry David Thoreau once wrote, "I went to the woods because I wished to live deliberately, to front only the essential facts of life, and see if I could not learn what it had to teach, and not, when I came to die, discover that I had not lived. . . . I did not wish to live what was not life; living is so dear. . . . I wanted to live deep and suck out all the marrow of life . . . to know it by experience . . ." While times have certainly changed since Thoreau chose to live and learn so deliberately in the woods, still the best education that can be given to children is one that teaches living and learning as deliberately—not just in the woods and wildernesses of the world, but, as well, in the cities and suburbs of everyday life. This is an education that offers enlightenment and understanding through self-discovery, encourages the development of self-confidence, and provides the challenges

necessary to build self-reliance.

Such an education cannot be taught only through books, for it is not built upon words. Such an education cannot be easily defined, for it must be discovered, oftentimes through trial and error. Such an education is, amazingly, both immaterial and useful. Such an education is the end product of great challenges that are mastered and great risks that are taken. Such an education comes out of understandings that are grasped on the way up the mountain of life or tripped over on the road less traveled. Such an education is comprised of the postcards that are accumulated on the journey through life—a lifelong journey that has no destination and no end. The foremost task of such an education is to ensure the survival as well as the preservation of each child's endless curiosity, tenacity for discovery, tender compassion for all living things, and, most of all, an inexhaustible interest in what makes the world, and the people in it, tick.

Some Ways to Educate with Nature

1. Discourage specialization or the development of mastery in just one area of interest. As writer Robert A. Heinlein once suggested, "A human being should be able to change a diaper, plan an invasion, butcher a hog, conn a ship, design a building, write a sonnet, balance accounts, build a wall, set a bone, comfort the dying, take orders, give orders, cooperate, act alone, pitch manure, solve equations, analyze a new problem, program a computer, cook a tasty meal, fight efficiently, die gallantly. Specialization is for insects." Remember, it is better to know a little about a lot than it is to know a lot about a little.

2. Support 4-H clubs—which stand for "head, heart, health, and hands"—and other organizations that combine friendships and fun with discipline and the development of responsibility.

3. Make sure school programs that include environmental education as part of the curriculum present both sides of vital issues. For example, teachers need to let children know that while recycling is a good thing, it often requires additional gasoline and continues polluting the atmosphere from the extra trucks that are needed to haul the recyclables. As well, make sure school environmental projects are based on scientific evidence and issues that children really are concerned about rather than are used to promote political activism.

4. Hire children to help you plant your garden, water your flowers and vegetables, prune your trees, build birdhouses, or fill your bird feeders. At the same time, teach them about the things they are tending.

5. Form a neighborhood "hike-pool" in which parents volunteer to take groups of kids out on nearby hiking trails on weekend afternoons.

6. Encourage your family members, partner, or housemates

to learn something new every day. Then set aside time to listen to and share what each has learned—perhaps during a communal meal or before settling down to watch a television show or video.

7. Be curious. Be open, receptive, and perceptive. Experiment. Observe. Ask. Touch. Taste. Smell. Find out how something works. Challenge yourself. Set a goal to run one mile further than you have ever run or to climb a mountain you have never climbed before. Know your limits, then surpass them.

8. Spend twenty-four hours alone in nature. Find out what you can learn about yourself and the world you live in by living more in it.

9. Learn how to act more responsibly with respect to people and the natural environment. Teach others what you learn through your actions.

10. Heal the wounds that have been inflicted upon the planet. Heal yourself. Heal those you love. Then reconnect with the planet, with yourself, and with those you love in life-enhancing ways.

For we are no longer isolated,
standing like starry visitors on a mountain-top,
surveying life from the outside;
but are on a level with and part and parcel of it;
and if the mystery of life daily deepens,
it is because we view it more closely
and with clearer vision.

—Argentine-born English writer
W. H. Hudson

Do you have an idea for a "Natural Act"? If so, the author would love to hear from you. Your "Natural Act" might even be included in a future book. Jot your "Natural Act" on a piece of paper and send it with your name, address, and daytime phone number to

Amy E. Dean
c/o M. Evans and Company, Inc.
216 East 49th Street
New York, NY 10017